This guide presents a variety of recommendations for improving indoor air quality in residential buildings through controlled mechanical ventilation.

These recommendations are not intended to apply to every conceivable situation but are intended to illustrate principles of best practice.

September 2006

WRITTEN BY: ARMIN RUDD

This guide was significantly enhanced by the thoughtful review of the following individuals whose input is greatly appreciated:

EEBA Project Manager: Ed VonThoma, VONED

EEBA Technical Committee: Farrel Beddome, Panasonic; Gord Cooke, Air Solutions; Steve Johnson, Andersen Windows; and Jim Larsen, Cardinal Glass

Terry Brennan, Camroden Associates
Stephanie Finegan, Building Science Corp.
Randy Folts, SelectBuild
Daniel Forest, Venmar Ventilation
Joseph Lstiburek, Building Science Corp.
Jeff Medanich, McStain Neighborhoods
Bernie Pallardy, Del Webb
Kelly Parker, Guaranteed Watt Savers
Betsy Pettit, Building Science Corp.
Don Stevens, Stevens Associates
Justin Wilson, Building Performance Solutions

Ventilation Guide

Building Science Press Inc.
70 Main Street
Westford, MA 01886

ISBN 0-9755127-6-5

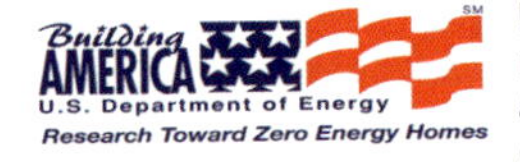

OFFICE OF BUILDING TECHNOLOGY STATE AND COMMUNITY PROGRAMS
OFFICE OF ENERGY EFFICIENCY AND RENEWABLE ENERGY
U.S. DEPARTMENT OF ENERGY

Book design and illustrations: Stephanie Finegan

Table of Contents

1. Principles

1.1 Why ventilate?

The primary purpose of mechanical ventilation is to provide a controlled exchange of more polluted inside air for less polluted outside air.

1.1.1 Health

Common residential indoor pollutants include excessive moisture, body odors, cooking emissions, volatile organic compounds (VOC's), combustion products, radon, pesticides, dust particles, virus, and bacteria. All of these are known to affect human health and/or comfort. Documented increases in asthma may be related to indoor pollutants (www.lungusa.org).

Excessive moisture is associated with mold and dust mites which can cause allergic and asthmatic reactions. Body odors are a nuisance, but not dangerous. Emissions from cooking, including nitrous oxides and grease should be vented at the source to avoid polluting the indoor air. Many VOC's, including formaldehyde, come from building materials, finishes, carpets, vinyl, adhesives, furnishings, and cleaning products. The degree to which VOC's are detected by people and affect human health, at concentrations usually found in residential indoor environments, is mostly unknown, and widely varies with each individual.

Combustion products from space or water heating appliances should never mix with the indoor environment. Sealed-combustion, direct-vent, and power vented appliances are widely available and should always be used whenever located inside the conditioned space.

Radon—a naturally occurring soil gas found in some areas of the country more than others—should be mitigated by a dedicated passive or active system separate from the local or whole-house ventilation systems discussed in this document. Local code and USEPA Guidelines should be followed.

Pesticides have clear health warnings for those coming into direct contact with them, but little is known about the gas phase effect in the indoor environment. It is best to use pesticides only as necessary, and indoor exposure pathways to treated soil should be sealed off.

Smoking, keeping pets indoors, hobbies, crafts, intensive cleaning and repainting or refinishing activities are in a class by themselves in terms of the production of indoor pollutants and the resultant health effect. In these cases, the occupants will have to use their own best judgments to reduce harmful exposures.

1.1.2 Comfort

Offensive odors, stale air, and stuffiness lead to occupant discomfort. The perception of comfort as it relates to indoor air quality varies greatly with each individual, with time, and with the setting. Some people are simply more sensitive to odors, and to different odors, than others. People also adapt to odors differently over time and may be more or less tolerant of odors in different environments.

1.1.3 Durability

Moisture affects both health and building durability. Once fire and structural requirements are met, moisture should be considered the number one enemy of a house. Durability with respect to moisture can be managed to a large extent through controlled mechanical ventilation when outside air is drier than inside air, but ventilation cannot be substituted for proper building design as follows:

The building:

- Draining rain off the structure, to the ground, and away from the building.
- Providing a building enclosure that can dry both to the inside and outside should it get wet.

The systems:

- Preventing excessive mechanical depressurization (greater than 5 Pascal) of occupied spaces and framing cavities (chases, soffits, interior walls).
- Preventing interior pressure differentials between closed rooms and the common area of more than 3 Pascal.
- Installing mechanical dehumidification separate from cooling for hot, humid climates.

2. How should we ventilate?

2.1 Source Control First

Even before a ventilation system design is considered, careful attention should be given to the following areas to preclude or control major sources of indoor air contamination:

- Use sealed combustion heating appliances if located inside conditioned space.
- Provide an airtight separation between attached garages and living spaces.
- Assure dry basements, crawlspaces, and walls.
- Use pesticides and cleaning agents wisely and store them safely.
- Choose building materials and finishes that are known to reduce pollutant emissions. Information about such products can be found at www.greenguard.org, www.chps.net, and www.usgbc.org.

2.2 Local Exhaust:

Concentrated Pollutant Removal

Bathrooms, kitchens, and laundry rooms are places where pollutants are generated in high concentration. When these areas are being used, an exhaust fan should be active to exhaust pollutants directly to the outdoors before they can negatively impact air quality elsewhere in the home. Typically, this is done with individual, surface-mounted, bathroom exhaust fans and a kitchen range hood. However, it can also be done with a single, remotely-mounted, exhaust fan or heat recovery ventilator pulling from multiple locations at once.

Occupancy sensors, humidity sensors, and light switch interlocks have all been used to try to influence the appropriate operation of local exhaust fans, but homeowner education remains the most useful tool. Builders should purposefully educate homebuyers, explaining that their participation in any truly successful indoor air quality strategy certainly includes operation of kitchen, bathroom and laundry exhaust fans when those spaces are being used. Especially in bathrooms, available time delay controllers which keep the exhaust fan energized for a time after the room is vacated are beneficial. Kitchen exhaust should always be active while the range or oven appliance is operating because they spill all of the combustion pollutants into the indoor environment. Also, assure that clothes dryer exhaust air is not too restricted and that it goes directly to outdoors.

2.3 Whole-House Ventilation:

Dilution for the Balance of General Pollutants

After local exhaust is applied to remove concentrated pollutants at their source, one must deal with the general body of pollutants that are distributed throughout the home, usually at a much lower concentration. Since these pollutants are dispersed, there is no practical way to capture and exhaust them as in a bathroom or over a kitchen cook-top. Rather, we use dilution to reduce the dispersed pollutants in the home. This is known as whole-house ventilation, where fresher outside air is distributed in a controlled manner to dilute more polluted inside air. Whole house ventilation can be operated continuously at a lower rate, or intermittently at a higher rate.

2.3.1 How much is enough?

While whole-house ventilation is needed for improved indoor air quality, ventilation rates that are higher than needed will waste energy. In dry climates, and during wintertime in cold climates, high ventilation rates can cause the house to be too dry, causing occupants to want humidification equipment that may not have been necessary if the ventilation rate were lower. Likewise, in hot, humid climates, high ventilation rates can increase indoor humidity beyond the latent capacity of the cooling system, requiring supplemental dehumidification.

Conversely, whole-house ventilation rates that are too low may result in odor discomfort and moisture problems. Medical health data is almost non-existent related to residential ventilation. In homebuilding practice, residential ventilation systems are likely deemed effective as long as occupants are satisfied with odor removal and moisture concerns do not arise.

There are some codes that require residential ventilation, most notably the states of Washington and Minnesota, the federal HUD Code for manufactured housing, and the National Building Code of Canada. Otherwise, standards and best practice guidelines can be applied.

2.3.1.1 ASHRAE Standard 62.2

In 2003, the American Society of Heating, Refrigeration, and Air Conditioning Engineers (www.ashrae.org) published the first national standard specifically for residential ventilation: Standard 62.2, **Ventilation and Acceptable Indoor Air Quality in Low-Rise Residential Buildings**. The standard was developed as a consensus among a committee with experience in the ventilation field. Right now, there is no whole-house distribution requirement in the standard. That means, for example, that a house with a single local exhaust fan in the master bathroom gets the same whole-house ventilation performance credit as a fully ducted ventilation system.

The ventilation rate established by Std. 62.2 is based on two factors—the number of bedrooms and the conditioned floor area. The bedroom part amounts to 7.5 ft^3 per minute (cfm) times the number of bedrooms+1. The floor area part (in cfm) amounts to one percent of the conditioned floor area. These two parts are added together. For example, a 3 bedroom, 2,000 ft^2 house would require 50 cfm, calculated as: $(7.5)(3+1) + (2000)(0.01) = 30 + 20$.

The whole-house ventilation can be continuous or intermittent. If intermittent, the full effective amount of ventilation must be provided within a three hour period or a higher ventilation rate must be used based on an effectiveness factor. For example, if the continuous ventilation flow rate was to be 50 cfm, the ventilation flow rate for a system operating one-third of the time (for example: 20 minutes per hour, or 60 minutes per three hours) would have to be 150 cfm. Ventilation fans that are not remotely mounted must also comply with a sound level requirement of 1 sone or less. That is quiet enough that background noise such as normal talking can mask the sound of the fan.

For local ventilation, a minimum of 50 cfm exhaust capacity is required per bathroom if operated intermittently, or 20 cfm if operated continuously. In kitchens, a minimum of 100 cfm exhaust capacity is required if run intermittently or enough exhaust to provide at least 5 air changes per hour if run constantly. If toilet rooms or utility rooms are not already served by an exhaust fan or dryer exhaust, there must be an operable ventilation opening to outdoors having an area that is at least 4% of the floor area but not less than 1.5 ft^2.

2.3.1.2 Home Ventilating Institute

The Home Ventilating Institute (HVI) is an industry organization whose members include HVAC system manufacturers and distributors (www.hvi.org). Members have their ventilation equipment tested and certified by the HVI according to standards adopted by the organization. For whole house ventilation, HVI recommends a minimum of 0.35 air changes per hour, or a rule-of-thumb of 5 cfm per 100 ft^2 of conditioned floor area. The rule-of-thumb method yields about two times the ASHRAE Standard 62.2 rate.

For local exhaust in bathrooms, HVI recommends 1 cfm per ft^2 or a code minimum of 50 cfm for bathrooms 100 ft^2 or less. For larger bathrooms, a fixture count is used. A toilet, shower, and bath tub require 50 cfm, while a jetted tub requires 100 cfm. A minimum three-fourths inch door undercut is recommended to aid in free exhaust flow, as is a delay timer to continue fan operation for at least 20 minutes after occupancy. For kitchen range hoods, HVI recommends 100 cfm per lineal foot for wall-mounted hoods and 150 cfm per lineal foot for island hoods.

2.3.1.3 Best practice guidelines based on homebuilding experience

Over the past two decades, hundreds of thousands of homes, both site built and manufactured homes, have been built throughout the U.S. and Canada with varying amounts of whole house mechanical ventilation. Research has shown that these homes have better perceived air quality and fewer moisture problems than homes that do not have whole house ventilation.

Over the last decade, some U.S. production homebuilders in partnership with the U.S. Department of Energy's Building America Program (www.buildingamerica.gov) have been on the leading edge of producing homes with low risk, and high durability, comfort, and efficiency using the systems approach. These homes include whole-house mechanical ventilation. Experience with over 100,000 such homes having central-fan-integrated supply ventilation with fan and damper cycling has been successful. Those homes have roughly 50 to 60 percent of the ventilation rate required by ASHRAE Standard 62.2. The lack of complaints by occupants indicates that the systems are working to provide indoor air quality acceptable to the occupants. Since these homes have effective whole-house distribution of ventilation air, it may be that effective ventilation air distribution can substitute for higher, non-distributed ventilation flow rates. Local codes must be met first. If there is no local ventilation code, then in our practice we recommend this level of whole house ventilation, along with enough available capacity to meet the ASHRAE 62.2 Standard upon occupant demand.

2.3.2 Combined infiltration and mechanical ventilation air change effects

Building enclosure air-tightening, in conjunction with controlled mechanical ventilation, should be substituted for the random infiltration of leaky buildings. This will help improve building performance and occupant satisfaction, including occupant health and comfort, moisture control, and energy efficiency. Space conditioning system size can usually be reduced because the high peaks of weather induced infiltration are controlled.

Random building air leakage through the enclosure, commonly called natural air infiltration, is completely dependent on environmental conditions. Natural infiltration can range from large to-almost non-existent, and one never knows when it will change.

When mechanical ventilation fan pressure forces are superimposed on natural infiltration forces, the result is not always additive as might seem logical. Supply and exhaust ventilation both tend to resist or control natural infiltration forces in tight houses. Balanced ventilation does not resist natural infiltration and tends to sum with it. In other words, for the same building enclosure, a balanced ventilation system will tend to allow more infiltration and yield a higher net air exchange rate than a supply or exhaust ventilation system.

2.3.3 Use of windows

Operable windows are needed for emergency egress, and can be useful for enjoying mild outdoor conditions, or for an occasional major airing-out due to some pollutant event, but they are not a reliable means of providing ventilation in modern homes. Opening and closing windows is not an effective strategy for whole house ventilation. Air flow through operable windows constantly changes based on environmental conditions of wind speed, wind direction, and indoor-outdoor temperature difference. The occupant has no way to judge how much air exchange is occurring. There is also no way to filter or condition that air. Homeowners may intuitively know this. According to a ventilation survey conducted by the California Air Resources Board on 1,500 new homes in California, few people use windows frequently. Top reasons for not using windows were: security, no-one at home, keeping noise out, too drafty, wind and rain, dust, dirt, pollen, and insects. Therefore, mechanical whole-house ventilation is needed regardless of available operable windows.

3. Types of Whole-House Ventilation Systems

Low air leakage is a requirement for an effective and energy efficient central space conditioning system or ventilation system. The entire air system must be substantially airtight, including all ducts, dampers, fittings, and the fan cabinet itself. If the system is leaky, this will defeat the purpose of intentionally sizing ducts and inlets/outlets to provide a controlled amount of air flow to individual spaces. Air leakage of less than five percent of the rated fan flow is recommended. For example, a central air handler with a flow of 1,000 cfm should have less than 50 cfm of air leakage.

3.1 Supply

There are several reasons to consider supply whole-house ventilation systems. These include:

- Introducing the outdoor air from a known source
- Treating the incoming air before it is distributed
- Minimizing the problem of pulling humid air into the building assemblies in humid summer regions
- Minimizing combustion appliance backdrafting potential

Supply ventilation systems draw outside air from a known location and deliver it to the interior living space. This known location should be selected to maximize the ventilation air quality. The air can be treated before distribution to the living space (heated, cooled, dehumidified, filtered, cleaned). If supply ventilation air is not pre-treated, it should be mixed with recirculated indoor air to mitigate discomfort effects of the outside air. Supply ventilation will tend to pressurize an interior space relative to the outdoors, causing inside air to be forced out through leak sites (cracks, holes, etc.) located randomly in the building enclosure. In warm, humid climates, this strategy minimizes moisture entry into the building enclosure from outdoors. Additional care should be taken with building enclosure design and workmanship when using supply ventilation in climates with very cold winters.

In some cold climate houses, depending on the interior humidity level, the building envelope design, the quality of workmanship, and the supply ventilation flow rate, exhaust capacity may be advisable to balance supply ventilation air to avoid continuous pressurization of the building. The following practices are recommended to avoid potential problems with supply ventilation in cold climate buildings:

1. Construct the building enclosure to dry to the exterior (vapor open to the outside) and/or use insulated sheathing to control the temperature of condensing surfaces.
2. Construct the building enclosure to avoid excessive air leakage that could transport interior moisture into the insulated cavities.
3. Control the indoor relative humidity in wintertime below 35% RH.

In any climate, interior humidity control is important to reduce condensation potential. Areas of high moisture generation such as kitchens, baths, and laundries, should be exhausted at the source. When outdoor conditions are right, whole-house ventilation then serves to dilute remaining interior moisture with drier outdoor air.

3.1.1 Integrated: Central-Fan-Integrated Supply System

Ventilation systems that provide ventilation air through a duct that extends from outdoors to the return air side of a central heating and cooling air distribution fan achieve full distribution of ventilation air using already existing ducts. These systems only provide ventilation air when the central system fan is operating, therefore, during mild outdoor conditions, the central fan may not be activated by the thermostat for long periods of time. Whole-house ventilation can be provided by:

1) running the central fan continuously;

2) operating the fan with a separate timer that disregards the fan's operation due to thermostat demand; or

3) operating the central fan with a fan cycling controller that takes into account prior fan operation due to heating and cooling and assures a minimum duty cycle.

Options 1 and 2 are less energy efficient, can shorten the life of the central fan motor due to more frequent operation, and are more likely to cause moisture problems in humid climates. Option 3 is the best practice solution because it allows operating the fan only as needed to meet the specified minimum duty cycle, and it helps avoid moisture re-evaporation immediately after a cooling cycle. All three options provide good distribution of ventilation air during stagnant periods when there is no thermostat demand to circulate air for purposes of heating or cooling. Use of electronically commutated motors (ECM) rather than permanent split capacitor (PSC) motors can reduce fan energy consumption by one-half or more. For best practice, a motorized outside air damper should also be added to limit outside air introduction to a maximum regardless of how long the fan operates.

Figure 1

Furnace mixed air return temperature

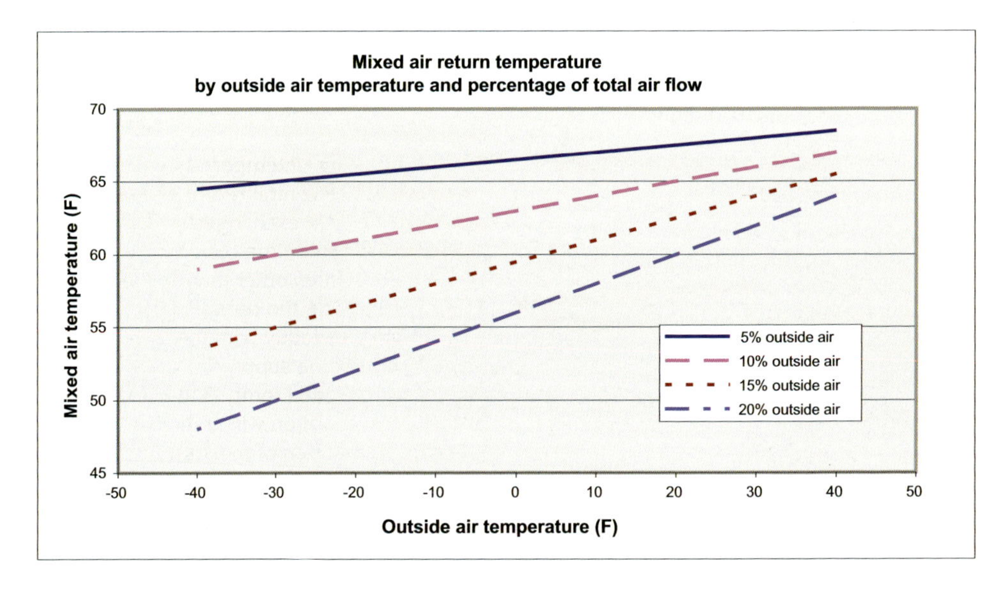

Because the operational time of the central system fan is increased, the importance of sealing and insulating the ducts is increased. Even better is when the entire air distribution system is located inside conditioned space.

Central fan operation can also provide improved temperature and humidity comfort control in conditioned spaces. Thermostats are typically located in a central area and are usually expected to also serve closed rooms, and often more than one floor level. Temperature conditions can vary widely between the thermostat location and the outermost areas of the home—mostly due to solar heat gain. Central fan operation tends to average the overall space conditions via whole-house air mixing. This can also improve the performance of dehumidification, filtration, and air cleaning equipment.

Limit the outside air flow to keep the mixed air return temperature above 55 °F, or follow the furnace manufacturer's requirements. Figure 1 gives the mixed air return temperature at various outside air temperatures and outside air flow percentages. For example, with 10% outside air (100 cfm outside air to a 1,000 cfm air handler), and an outside temperature of -20 °F, the mixed air temperature at the air handler return will be 61 °F.

3.1.2 Stand-alone: Separate Supply Fan and Full Duct System

This system includes a separate duct system dedicated for supplying ventilation air. This may be a good option in houses without central air handlers. Multi-point supply ventilation systems should supply ventilation air to the common area and to all closable rooms. Care must be taken to avoid cool air discomfort during cold wintertime conditions, therefore, tempering of the ventilation air with 3 parts indoor recirculation air is required.

3.1.3 Combination: Separate Supply Fan Integrated with a Central Duct System

To reduce the cost of ventilation ductwork, it may seem desirable to supply ventilation air with a separate supply fan into the return or supply side of a central duct system. However, in hot humid climates this should never be done. In other climate regions there are a number of factors that must be considered first as follows:

- If the separate supply ventilation fan does not operate continuously, a motorized damper (well sealed, normally-closed, spring return) must be installed to eliminate duct leakage to outside that will occur when the central fan operates and the supply ventilation fan does not.
- If injecting into the central supply duct, then the separate supply fan must be designed to overcome the static pressure in the central supply duct when the central fan is operating.
- A minimum mixing ratio of 3 parts recirculation air to 1 part outside air should be used to temper the ventilation air.
    - Injecting untempered outside air into the central return can cause furnace heat exchangers to fail during cold wintertime conditions.
    - In climates other than hot humid, operating the central air handler fan constantly with the separate ventilation supply fan will provide the needed tempering but energy consumption will be higher and re-evaporation of moisture from cooling coils can reduce indoor humidity control.

This system will not achieve whole house distribution of ventilation air when the central fan is not operating. The relatively low ventilation supply air flow will follow the path of least resistance reaching only some of the duct outlets. Periodic operation of the central fan will provide whole-house distribution.

3.2 Exhaust

Exhaust whole-house ventilation systems expel inside air directly to outdoors, tending to depressurize the interior space relative to outdoors. Exhaust systems draw outdoor air from whatever building enclosure leaks create the path of least resistance. Passive air inlet vents can be provided, but these inlets have been shown to make only marginal improvement and may serve as either an inlet or an outlet, increasing natural infiltration.

During cold wintertime conditions, exhaust depressurization minimizes potential condensation problems in walls from excessive indoor moisture. Excessive depressurization in humid cooling climates invites moisture problems due to condensation of exterior moisture within the building enclosure. This is typically not a problem with small exhaust fans (30 to 90 cfm), but it can be a major problem with large exhaust fans.

With exhaust-only ventilation, it is not possible to filter or condition (heat/cool/dehumidify) the outside air before it enters the living space. The building enclosure becomes somewhat of a filter itself as sometimes evidenced by dust markings at baseboards on light carpets. It is not possible to know where the ventilation air comes from. Outside air that comes through garages, attics, crawlspaces, or other enclosure surfaces in contact with soil may be contaminated with gaseous pollutants, or particulates. Any combustion heating appliances within the conditioned space should be at least power vented, but preferably direct-vent, sealed combustion.

3.2.1 Fan with one room inlet

Single-point whole-house exhaust ventilation most commonly entails a high quality bath fan installed in a master bathroom, family bathroom, powder room, or laundry. In some cases, a dedicated fan will be installed in a ceiling location in the central area of the house. These fans are generally quiet, rated for continuous duty, and have low power draw. These fans are typically surface-mounted, inline, or remote-mounted. Some have multiple speeds to allow for double-duty as both the bath fan and the whole-house fan.

This system will generally not achieve whole-house distribution of ventilation air, especially to closed or distant rooms, since the replacement air will follow the path of least resistance flowing from the closest leak sites. Periodic operation of the central HVAC system fan at least 10 minutes per hour will provide whole-house distribution.

3.2.2 Fan with multiple room inlets

Multi-point exhaust ventilation systems should exhaust from all closable rooms and the common area. This gives the best distribution and the quietest airflow. These systems require separate ducts for distributing ventilation air only. To reduce ductwork and redundant exhaust equipment, multi-point exhaust systems sometimes exhaust only from bathrooms, leaving the air exchange of areas not directly connected to those bathrooms in question.

3.3 Supply/Exhaust Combination with net imbalance

Supply and exhaust whole-house ventilation systems may be mixed in combination such that the net result is an average air flow imbalance. An example application of this system could be central-fan-integrated supply ventilation along with continuous or intermittent exhaust. That system may be preferred in very cold climates to avoid continuous pressurization of the house. Another strategy could be to provide a lower background level of supply ventilation that can be augmented upon occupant demand with a higher level of exhaust ventilation to increase outside air exchange for a time.

3.4 Balanced

Balanced whole-house ventilation systems both exhaust and supply in roughly equal amounts. Inside air is exhausted to the outdoors and outside air is supplied indoors. Balanced ventilation, by definition, should not affect the pressure of an interior space relative to the outdoors. In reality the balance may never be perfect due to fluctuations in wind and stack pressures. In current practice, most balanced ventilation systems are of the heat recovery or energy recovery type, but they don't have to be. A balanced system can be made of any combination of the exhaust and supply ventilation systems described above. Balanced ventilation can be used effectively in any climate without reservation. Outside air should be taken from a known fresh air location and filtered before it enters the living space.

3.4.1 Single-point (one exhaust and one supply point)

Single-point balanced ventilation systems exhaust ventilation air from one location in the house and supply outdoor ventilation air to another location. These systems generally will not achieve whole-house distribution of ventilation air. Closed rooms without a ventilation supply or exhaust point generally will not receive adequate ventilation air unless by whole-house mixing from central system fan operation.

3.4.2 Multi-point (fully ducted)

Multi-point balanced ventilation systems should either supply or exhaust air from or to all closable rooms and the common area. These systems will achieve whole-house distribution of ventilation air and require separate ducts for distributing ventilation air only. There can be various combinations of multi-point and single-point supply and exhaust fans.

3.4.3 Heat recovery (HRV) or energy recovery (ERV) systems

Balanced ventilation systems with heat recovery operate the same as the balanced ventilation systems described above with the exception that a heat exchanger transfers some heat between the exhaust air stream and the outside air supply stream. No moisture is exchanged between the air streams. This means that in cold months, the heating load due to ventilation will be less, and in hot months, only the sensible cooling load due to ventilation will be less.

Balanced ventilation systems with energy recovery operate the same as the balanced ventilation systems described above with the exception that both heat and moisture are exchanged between the exhaust air stream and the outside air supply stream. This means that in cold, dry months, the heating load due to ventilation will be less, and the house interior moisture level will be higher than it otherwise would have been without energy recovery. In hot, humid months, the total cooling load (both sensible and latent) due to ventilation will be less. While less heat and moisture will come in from outdoors, an energy recovery ventilator can neither cool nor dehumidify the interior space. A good way to think of this is that the heat and moisture tend to remain on the side from which they came.

Any HRV/ERV that is connected to the central system supply side must have a motorized damper to keep air from flowing through the unit when the ventilator is off.

Connecting both the HRV/ERV inlet and outlet to the central system return and supply ducts, respectively, or connecting both the inlet and outlet of the HRV/ERV to the central return duct requires coincident operation of the central fan. Constant central fan operation is energy intensive and causes humidity problems in humid cooling climates due to re-evaporation of moisture from the cooling coil. Use of electronically commutated motors (ECM) rather than permanent split capacitor (PSC) motors can reduce fan energy consumption by one-half or more.

In humid summertime conditions, even with moisture (latent) recovery, the dew point of the ventilation supply air can be much higher than the cool central supply ducts. This can lead to condensation and mold in those ducts when the HRV/ERV is active but the central fan is not. This can be a problem whether the HRV/ERV ventilation air is injected into the supply or return of the central system. Injecting the ventilation air into the central return upstream of a relatively restrictive media filter installed at the inlet of the air handler makes it less likely that the ventilation air will flow into the cool central supply ducts.

It is better to provide a dedicated duct system independent of the central ducts. Such a dedicated duct system could be fully-ducted or minimally ducted with use of periodic operation of the central air handler for whole-house distribution.

In cold climates, HRV's and ERV's must be installed in conditioned or tempered spaces to minimize heat exchanger frosting and freezing of condensate, unless manufacturer's guidelines are followed for protecting against this.

4. Types of control strategies

All ventilation systems, both local and whole-house, need controls to operate effectively. The primary types of ventilation controls are time-based, speed-based, and sensor-based.

Simple time-based control is inexpensive and uncomplicated. Such a control may be the common bathroom crank delay timer that activates a fan for a given time then stops. With microprocessors, push-buttons and programming can increase the ease of use and flexibility, and eliminate the mechanical noise of the crank timer. Older style pop-up timers can be set to activate and deactivate a fan creating a sequentially programmed operation over the course of a day. With microprocessors, that programming can be made easier and more flexible, while eliminating the noise of electromechanical switching. A runtime fraction, or duty cycle can also be programmed to cause a certain amount of operation over a given time period.

Adding logic to simple time-based control increases the complexity but can also increase the value. For example, accounting for prior operation of ventilating or circulating fans due to occupant activation or activation by another device can be an advantage. It can reduce operating cost, reduce equipment noise and wear-and-tear, and improve overall system acceptability and performance.

Time-based control can be applied to motorized outside air dampers as well as to fans. A motorized damper in an outside air duct can be opened in conjunction with fan activation to assure a minimum amount of ventilation air introduction. The motorized damper can also be closed independent of fan operation to limit outside air introduction to a programmed maximum.

Fan speed controls allow adjustment of the motor speed which changes the ventilation flow rate. Some fan speed controllers allow multiple steps in speed control. For example, a lower speed may be selected for normal automatic operation and a higher speed may be manually activated as a "bump-up" speed for a time.

Sensors can be used to activate and deactivate ventilation equipment based on some measured condition. The most common measurements used in residential ventilation control are indoor relative humidity, occupancy, and outdoor temperature.

4.1 Time-based

The primary types of time-based ventilation controls are continuous and intermittent.

4.1.1 Continuous

Continuous operation is the simplest operation strategy. It requires nothing but a simple on/off switch. No electronics or sensors are needed for measuring some trigger condition and making a control decision. Continuous operation can be applied to supply or exhaust fans, HRV's and ERV's, or the central air handler, and ventilation rates are lower for continuous operation than for intermittent. Continuous operation can be encouraged if the system is quiet and efficient. If it is not, people tend to turn the equipment off and often forget to turn it back on.

Continuous operation also requires that the ventilation air flow rate be designed and setup properly once and for all. There is no variability to modify the ventilation amount up or down by changing the amount of time the system runs. Occupants may want that variability depending on their lifestyle and preferences.

4.1.2 Intermittent

Intermittent control of ventilation allows for a larger system that can normally be operated a fraction of the time but can be operated more depending on occupant preference or the level of indoor pollutants.

The minimum intermittent operation time is often set to protect against excessive indoor moisture buildup and odor complaints. If the occupant prefers more ventilation, the operation time can be increased.

4.1.3 Combination

A combination of continuous and intermittent time-based ventilation control can be used effectively. For example, a supply ventilation system with periodic operation of the central system fan may be used in conjunction with continuous single-point exhaust to provide the following benefits:

a) a known entry point for supply ventilation air that can be filtered and conditioned or tempered;

b) intermittent pressure balance reducing the amount of time the space is depressurized;

c) whole-house distribution of ventilation air, and mixing for improved thermal comfort.

4.2 Speed-based

Motor speed controllers can be used effectively to allow adjustment of the base rate of ventilation and to provide for an increase in ventilation rate upon demand. This increases system flexibility so that the same equipment can be applied in different situations that require different ventilation rates.

For example, an exhaust fan or HRV/ERV may have a standard rated flow of 150 cfm which may be 2 or 3 times the continuous ventilation air flow desired for a given house. A motor speed controller could be applied to reduce the ventilation air flow to 50 or 75 cfm continuous, but the occupant might have the option to press a button to increase it to full capacity for a time.

Compatibility of the fan motor and speed controller should be verified. Some combinations of fan motor design and speed controller design are not compatible and may cause an objectionable motor noise or early motor failure.

4.3 Sensor-based

Sensors are used to measure and provide feed-back to a logic controller which makes a control decision. Indoor relative humidity, occupancy, and outdoor temperature are most commonly used to control residential ventilation.

For example, if the relative humidity in a bathroom or other space rises too high, an exhaust fan may be activated to remove the excess moisture. In climates where outdoor conditions are usually drier than indoor conditions, indoor relative humidity measurement is sometimes used to activate ventilation to reduce indoor humidity.

An occupancy sensor placed in a bathroom can assure that local exhaust will be activated whenever the bathroom is being used. For added convenience, occupancy sensors and delay-off timers for bath fans can be combined. Occupancy sensors can also be used to turn off the ventilation system to save energy while the house is not occupied.

Some controllers use measurement of outdoor temperature to diminish or eliminate whole-house ventilation under certain extreme temperature conditions (low or high).

5. Influences on Ventilation System Decisions

Perception of overall value is an individual choice based on a number of factors. The most significant factors are likely to be performance, initial cost, operating cost, non-climatic factors and climatic factors.

5.1 Performance

From an engineering perspective, the best performing systems over all climates would range from a fully-ducted, balanced heat or energy recovery system that operated continuously, to an intermittent central-fan-integrated supply system with available exhaust capacity, to a single-point exhaust fan located in a closable room.

Performance is established by the effectiveness of the system to provide uniformly distributed ventilation air for the dilution of indoor pollutants. The ease and cost of operating and maintaining the system will then factor in to the occupants overall acceptance of use.

Acceptance of use by the occupants can make the difference between a ventilation system that is turned **ON** or left **OFF**. The following ventilation system problems can all defeat its acceptance:

- Noise
- Air flow or temperature discomfort
- Perception of excessive humidity or dryness
- Dirt/dust markings on carpets
- Dirt/dust accumulation on horizontal surfaces
- Inconvenient or expensive maintenance
- Perception that the system costs too much to operate relative to the perceived benefit.

The best system will operate unobtrusively enough in the background so that its operation becomes accepted as normal.

5.2 Initial Cost

For production new construction, the initial cost of purchased equipment and installation is often the most significant factor affecting ventilation system choices. Initial cost will increase from single-point exhaust, to central-fan-integrated supply with or without central fan cycling and/or damper cycling, to multi-point exhaust or supply, to HRV/ERV systems whether fully ducted or not.

After initial cost, the longer-term impacts on both the building and occupants are then factored in to create an overall perceived value as the basis of choosing a ventilation system design.

For example, upgrading from a standard builder-grade bathroom exhaust fan to a quiet, higher-quality bathroom exhaust fan is clearly the lowest initial cost. But to gain whole-house distribution, without adding more fans or ductwork, a value decision might be made to add a fan cycling control on the central system fan to assure periodic mixing throughout the house. The initial cost of a fan cycling control is about the same as one additional upgraded bathroom fan or one additional duct run (inlet or supply).

Continuing with that example, to help avoid problems sometimes associated with exhaust ventilation, like the unknown source of unfiltered/unconditioned air and dust marking on carpets, an additional value decision might be made to use supply ventilation with an outside air duct to the central system return for at least part of the time.

5.3 Operating Cost

The cost to operate different ventilation systems is less important to production homebuilders than to homeowners. Annual energy use simulations for cities across the United States show that, taking the single-point exhaust system as a reference, the difference in total HVAC operating cost between a full range of ventilation systems for a given city is generally less than $100 per year.

5.4 Non-Climatic Influences

The following are ventilation system decision factors which must be considered regardless of climate:

a) Where is the ventilation air coming from and what pollutants might be coming in with it?

b) Should the ventilation air be filtered and conditioned or tempered before delivery to the living space?

c) How well is the ventilation air distributed throughout the living space?

d) Will depressurization contribute to a combustion safety problem?

e) Will filters be easily accessible and are they likely to be changed or cleaned?

f) Will the electrical and mechanical parts last, and are they likely to be repaired or replaced if they fail?

g) Will noise be an annoyance?

h) Will uncomfortable air temperature or air movement be an annoyance?

5.5 Climatic Influences

Buildings can be constructed and operated such that climate would not be a critical factor in the choice of a ventilation system. For example, supply-only ventilation can work well in a cold climate as long as:

1) the building enclosure can dry to the exterior (vapor open) and insulated sheathing controls the temperature of condensing surfaces;

2) the building enclosure air leakage is uniformly low; and

3) wintertime indoor relative humidity is limited to 35 percent.

Likewise, exhaust-only ventilation can work in warm, humid climates as long as:

1) no water vapor retarder impedes summertime drying to the interior;

2) the cooling set point temperature is not generally below 75 °F; and

3) exterior claddings are applied over an effective drainage layer.

If the enclosure design and operating conditions cannot be controlled, or are unknown, then some basic principles regarding climate and ventilation system design apply:

- Balanced ventilation works well in any climate because it does not alter the house air pressure with respect to outdoors. In that case, air is not mechanically forced to move in either direction through the building enclosure.
- From a very general, top-level perspective, exhaust whole-house ventilation works best in colder climates. That is because generally drier outside air moves inward through the building enclosure and moist interior air is forced out through a duct. In this case, depressurization keeps moist interior air from moving outward where it could condense on cold surfaces.
- From the same very general, top-level perspective, supply ventilation works best in warm, humid climates. That is because generally drier indoor conditioned air moves outward through the building enclosure. In this case, pressurization keeps moist outdoor air from moving inward where it could condense on cool surfaces.
- In dry and mild climates, either supply or exhaust ventilation can be used equally from a moisture perspective.

6. Best practice system details by climate

This section is organized by climate with the expectation that most users will not read through each climate but rather focus on the climate where they build or live. For those who study the details for more than one climate, you will notice that many figures are duplicated with slight changes or no changes.

Within each climate, the ordering of best practice details are generally intended to range from the most recommended to the least recommended system based on performance, initial cost, operating cost, non-climatic influences, and climatic influences.

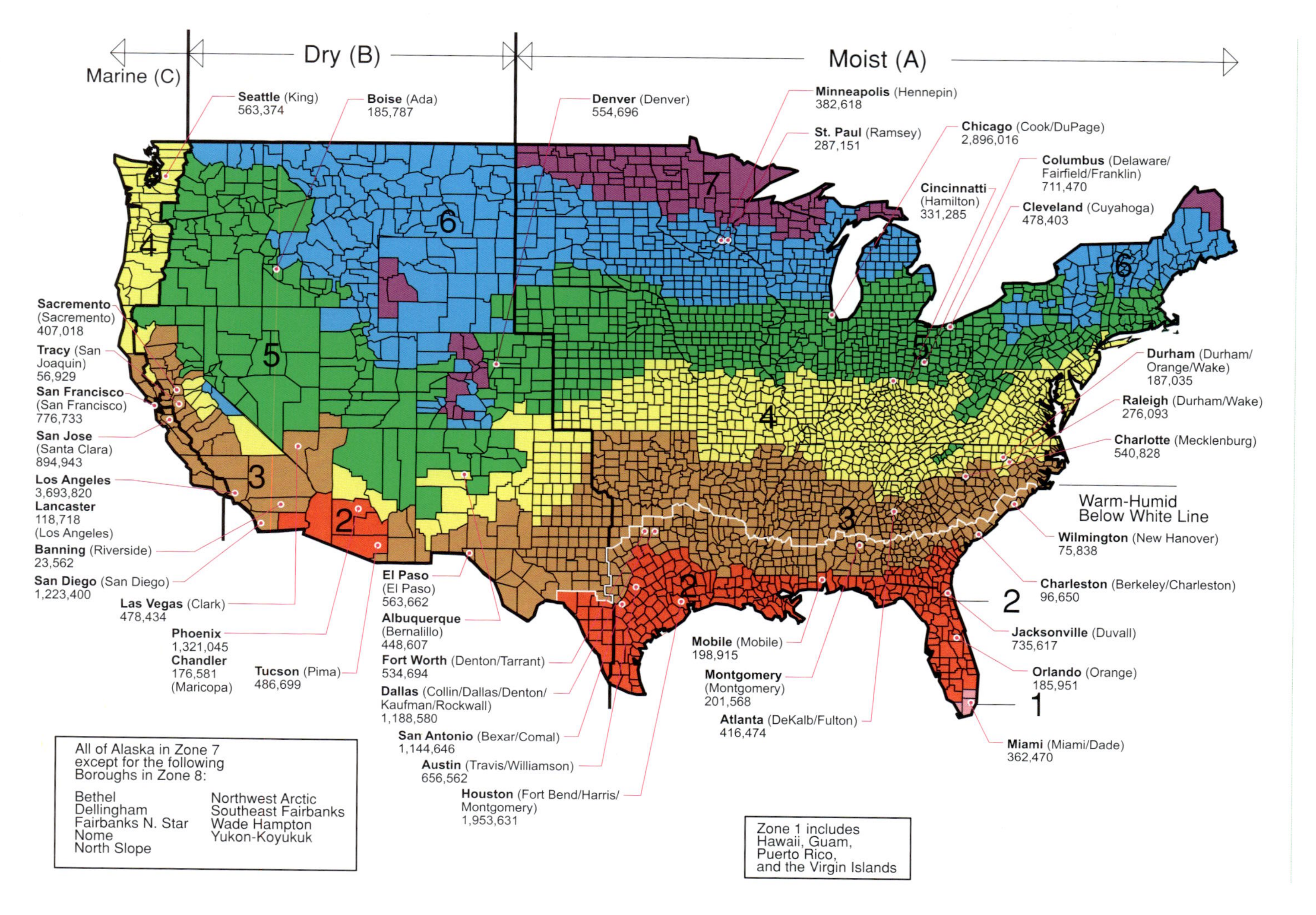
Marine (C)
Dry (B)
Moist (A)
Seattle (King)
563,374
Boise (Ada)
185,787
Denver (Denver)
554,696
Minneapolis (Hennepin)
382,618
St. Paul (Ramsey)
287,151
Chicago (Cook/DuPage)
2,896,016
Cincinnati
(Hamilton)
331,285
Columbus (Delaware/
Fairfield/Franklin)
711,470
Cleveland (Cuyahoga)
478,403
Durham (Durham/
Orange/Wake)
187,035
Raleigh (Durham/Wake)
276,093
Charlotte (Mecklenburg)
540,828
Warm-Humid
Below White Line
Wilmington (New Hanover)
75,838
Charleston (Berkeley/Charleston)
96,650
Jacksonville (Duvall)
735,617
Orlando (Orange)
185,951
Miami (Miami/Dade)
362,470
Sacremento
(Sacremento)
407,018
Tracy (San
Joaquin)
56,929
San Francisco
(San Francisco)
776,733
San Jose
(Santa Clara)
894,943
Los Angeles
3,693,820
Lancaster
118,718
(Los Angeles)
Banning (Riverside)
23,562
San Diego (San Diego)
1,223,400
Las Vegas (Clark)
478,434
Phoenix
1,321,045
Chandler
176,581
(Maricopa)
Tucson (Pima)
486,699
El Paso
(El Paso)
563,662
Albuquerque
(Bernalillo)
448,607
Fort Worth (Denton/Tarrant)
534,694
Dallas (Collin/Dallas/Denton/
Kaufman/Rockwall)
1,188,580
San Antonio (Bexar/Comal)
1,144,646
Austin (Travis/Williamson)
656,562
Houston (Fort Bend/Harris/
Montgomery)
1,953,631
Mobile (Mobile)
198,915
Montgomery
(Montgomery)
201,568
Atlanta (DeKalb/Fulton)
416,474
1
2
3
4
5
6
7
All of Alaska in Zone 7
except for the following
Boroughs in Zone 8:
Bethel
Dellingham
Fairbanks N. Star
Nome
North Slope
Northwest Arctic
Southeast Fairbanks
Wade Hampton
Yukon-Koyukuk
Zone 1 includes
Hawaii, Guam,
Puerto Rico,
and the Virgin Islands

Hot-Dry Climate

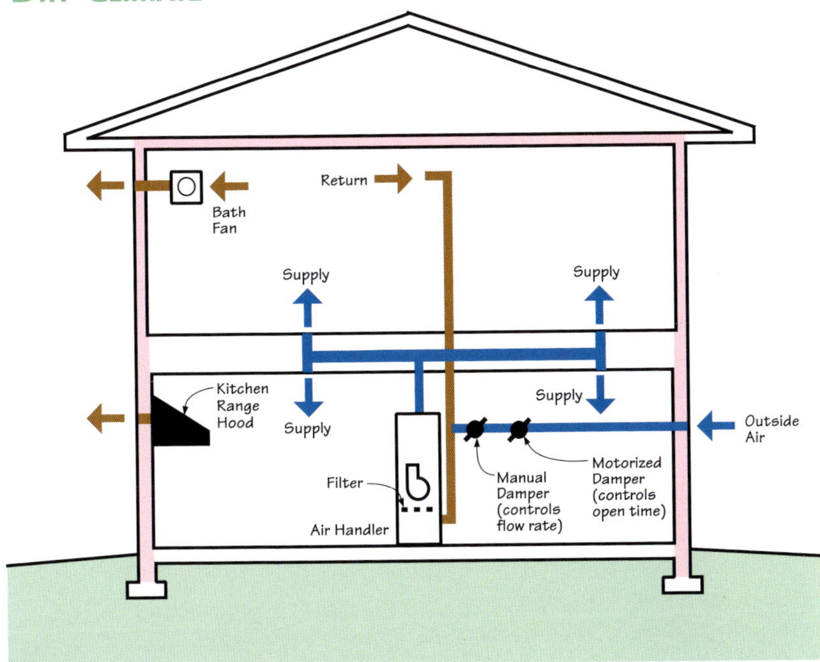

Figure 6.1.1

- Central-fan-integrated **supply** ventilation with available bathroom **exhaust**. Outside air intake through sidewall.
- Manual balancing damper in outside air duct allows adjustment of the flow rate.
- Periodic operation of the central air handler fan assures consistent ventilation air distribution and uniform air quality. It also reduces temperature variations between rooms.
- Optional motorized damper closes the opening to outside when the fan is off. With damper cycling control, outside air intake can be limited independently of how long the fan runs.
- Keeping all ducts inside insulated space provides the best performance. Sealed and well insulated ducts are next best.

Hot-Dry Climate

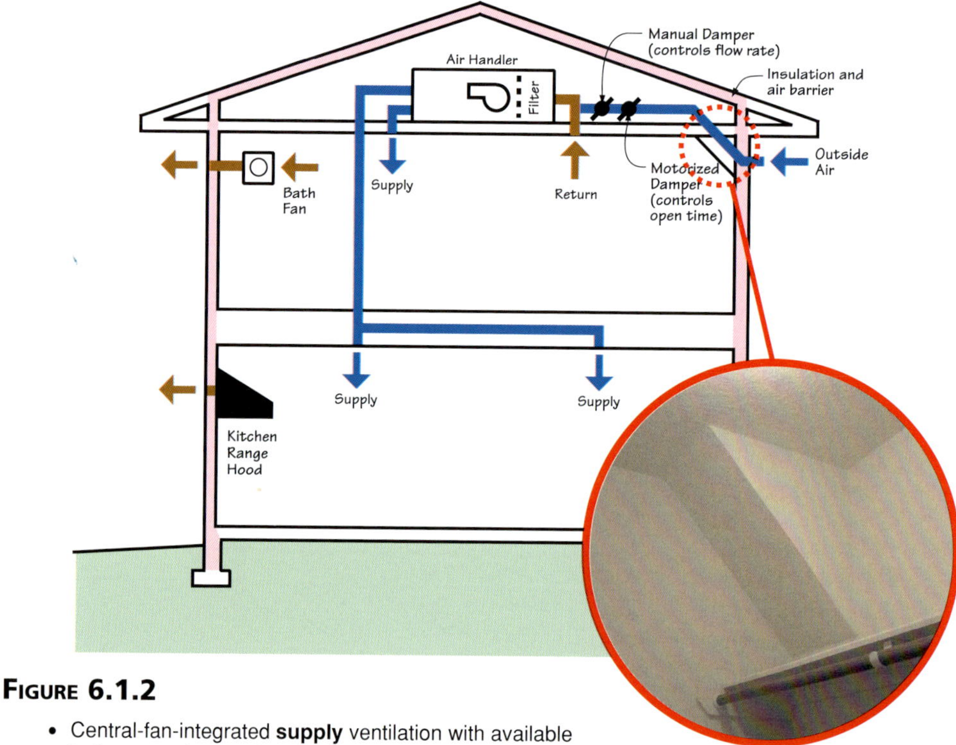

Figure 6.1.2

- Central-fan-integrated **supply** ventilation with available bathroom **exhaust**. Outside air intake through sidewall via angled fur-down in a closet (going through a gable end is just as good but avoid going through the roof).
- Manual balancing damper in outside air duct allows adjustment of the flow rate.
- Periodic operation of the central air handler fan assures consistent ventilation air distribution and uniform air quality. It also reduces temperature variations between rooms.
- Optional motorized damper closes the opening to outside when the fan is off. With damper cycling control, outside air intake can be limited independently of how long the fan runs.
- Keeping all ducts inside insulated space provides the best performance, such as permitted by the **unvented-cathedralized attic** shown above. Sealed and well insulated ducts are next best.

Hot-Dry Climate

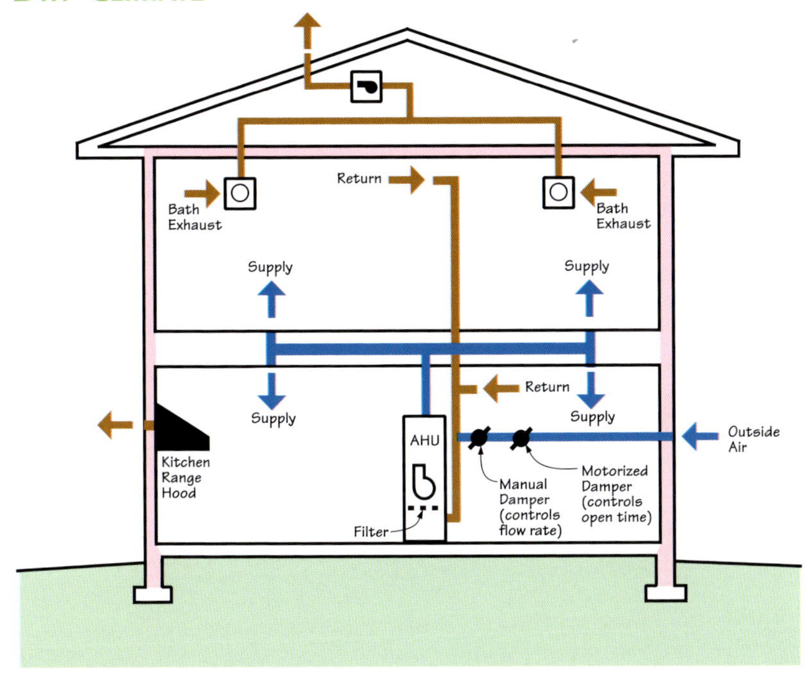

Figure 6.1.3

- Central-fan-integrated **supply** ventilation with multi-point **exhaust** from bathrooms. Outside air intake through sidewall.
- Occupant activated bathroom switches change the fan to high speed for a time.
- Manual balancing damper in outside air duct allows adjustment of the flow rate.
- Periodic operation of the central air handler fan assures consistent ventilation air distribution and uniform air quality. It also reduces temperature variations between rooms.
- Optional motorized damper closes the opening to outside when the fan is off. With damper cycling control, outside air intake can be limited independently of how long the fan runs.
- Keeping all ducts inside insulated space provides the best performance. Sealed and well insulated ducts are next best.

Hot-Dry Climate

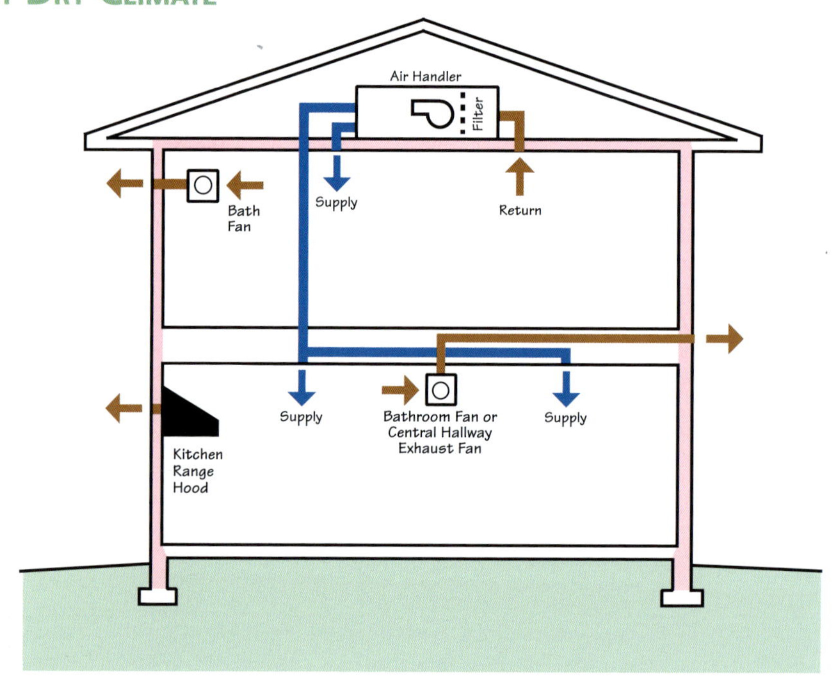

Figure 6.1.4

- Single-point **exhaust** ventilation, located preferably in a central hallway or common area. Preferably not located in a utility room off the garage.
- For improved air quality, periodic operation of the air handler fan provides whole-house mixing for distribution of ventilation air. It also reduces temperature variations between rooms.

Hot-Dry Climate

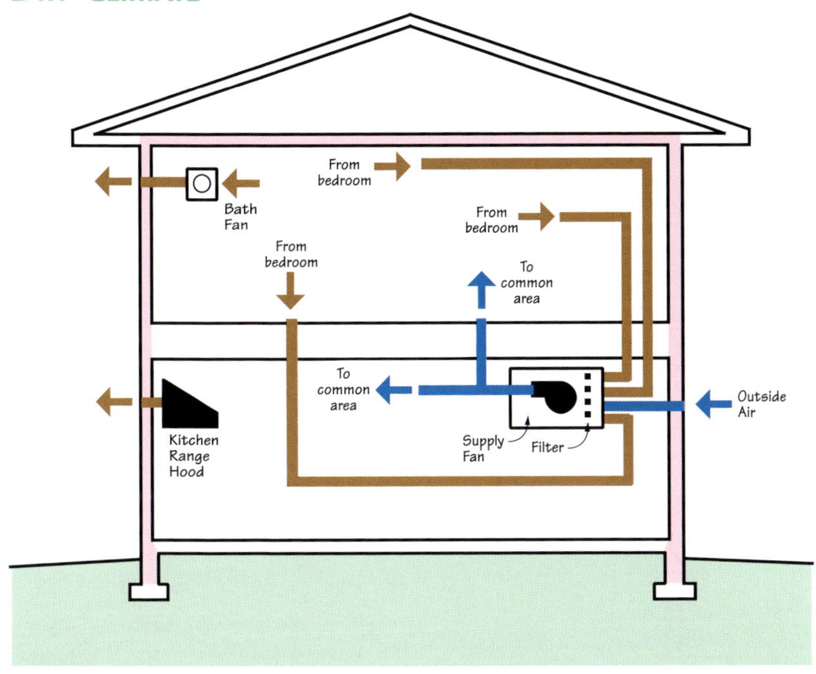

Figure 6.1.5

- A single- or multi-point **supply** ventilation system using a remote-mounted packaged fan with filter.
- This system mixes 1 part outside air with 3 parts inside air for tempering, then filters and delivers the mixed air to the common area.
- If single-point, the preferred recirculation air pickup should be from the master bedroom. This tends to cause ventilation air to flow back to the master bedroom through the common area. For improved air quality, periodic operation of the air handler fan provides whole-house mixing for distribution of ventilation air. It also reduces temperature variations between rooms.
- If multi-point, the recirculation air pickups should be from all bedrooms. This tends to cause ventilation air to flow back to the bedrooms through the common area. Alternately, the supply outlets could be in each room with the recirculation air pickup in the common area, however, care should be taken with supply outlets to avoid potential air discomfort over beds.
- The common area supplies should be high wall supplies to minimize any air flow discomfort.

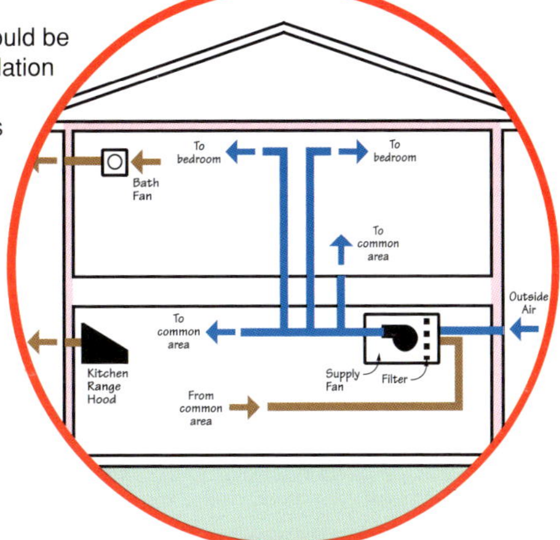

Hot-Dry Climate

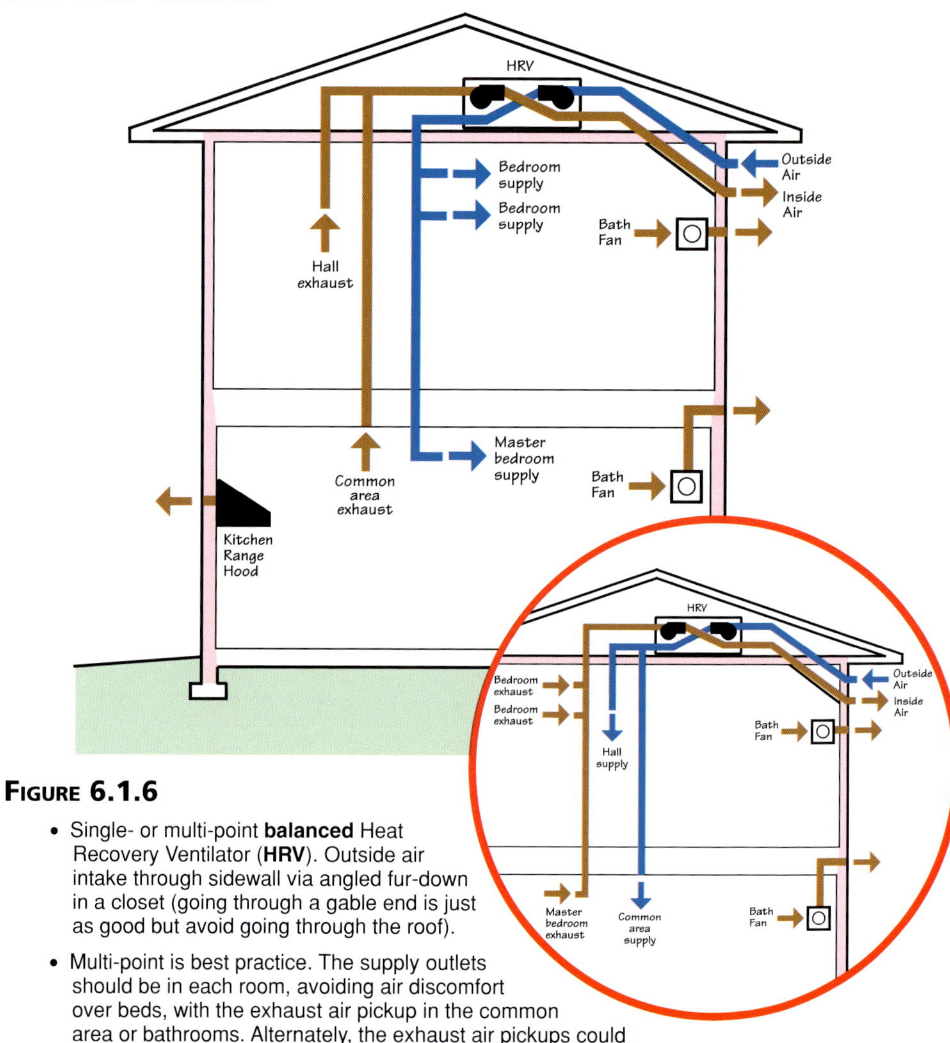

Figure 6.1.6

- Single- or multi-point **balanced** Heat Recovery Ventilator (**HRV**). Outside air intake through sidewall via angled fur-down in a closet (going through a gable end is just as good but avoid going through the roof).
- Multi-point is best practice. The supply outlets should be in each room, avoiding air discomfort over beds, with the exhaust air pickup in the common area or bathrooms. Alternately, the exhaust air pickups could be from all bedrooms, and the ventilation air supply could be to open common areas. This tends to cause the ventilation air to flow back to the bedrooms through the common area.
- If single-point, the preferred exhaust air pickup should be from the master bedroom, and the ventilation air supply should be in the open common area. This will tend to cause the ventilation air to flow back to the master bedroom through the common area. For improved air quality, periodic operation of the air handler fan provides whole-house mixing for distribution of ventilation air. It also reduces temperature variations between rooms.
- The outside air intake and exhaust outlet should be separated by about 10 feet or follow manufacturer's recommendation to avoid contamination of ventilation air.
- If the outside air intake and exhaust outlet are run through the eave soffit instead of through the sidewall, specific care must be taken to: 1) not crush the ventilation ducts between the wall top plate and the roof sheathing; 2) leave the full depth of ceiling insulation in place; and 3) fully terminate the ducts through the finished soffit to outdoors so that air flow is not restricted and leakage to/from the attic will not occur.

Hot-Dry Climate

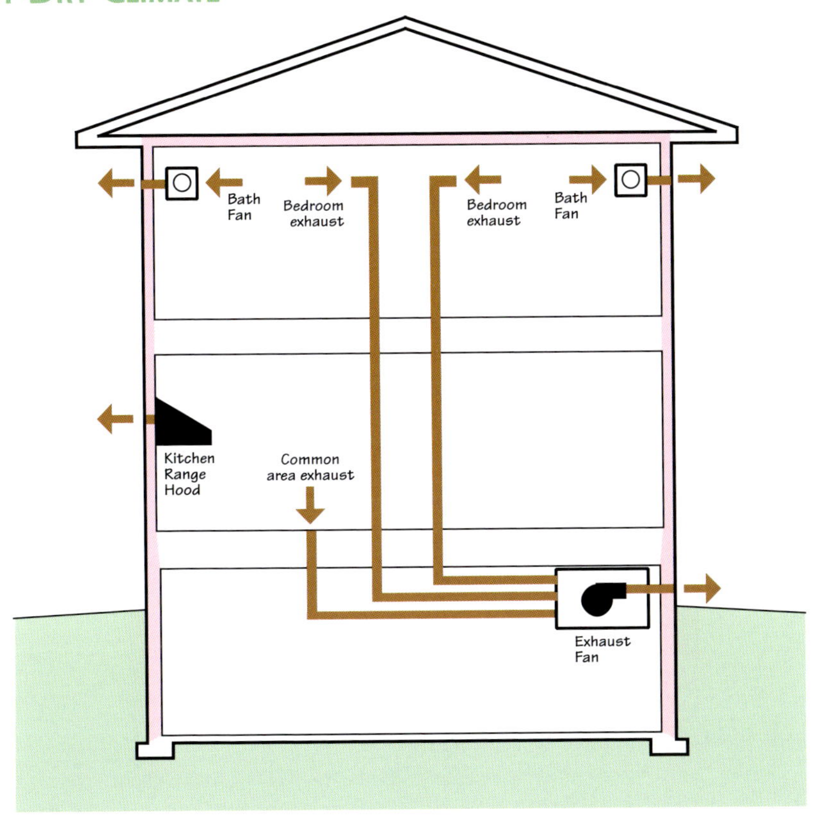

Figure 6.1.7

- Multi-point **exhaust** pulling air continuously from each bedroom and the common area.
- Best practice is to not locate natural draft combustion appliances in the indoor environment. As with all exhaust only ventilation systems, special attention should be given in order to prevent problems such as backdrafting of combustion appliances, intaking of soil gases, or intaking of air from an attached garage. The larger the exhaust air flow, the more the concern.

Hot-Dry Climate

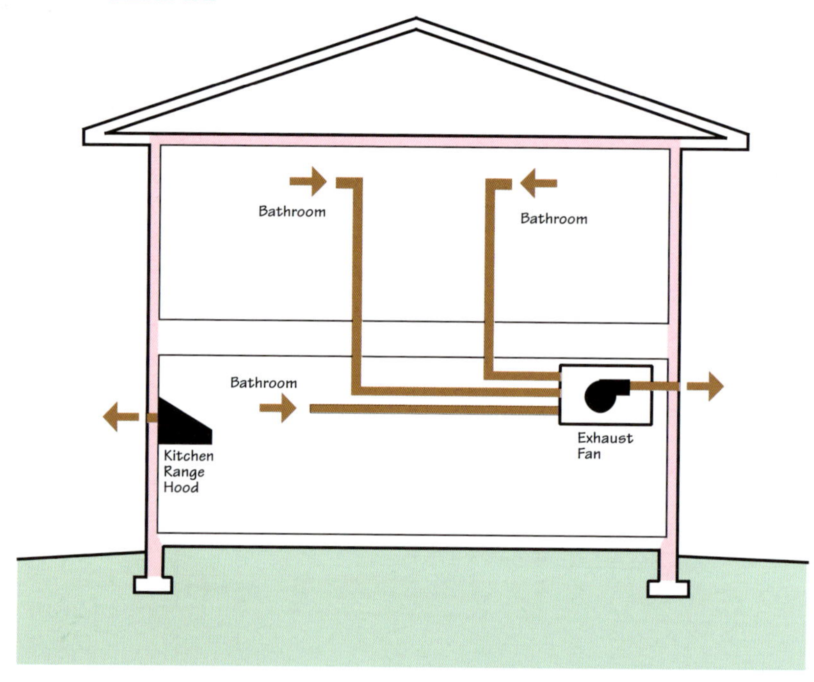

Figure 6.1.8

- Multi-point **exhaust** pulling air continuously from each bathroom.
- Occupant activated bathroom switches change the fan to high speed for a time.
- Best practice is to not locate natural draft combustion appliances in the indoor environment. As with all exhaust only ventilation systems, special attention should be given in order to prevent problems such as backdrafting of combustion appliances, intaking of soil gases, or intaking of air from an attached garage. The larger the exhaust air flow, the more the concern.

Hot-Humid Climate

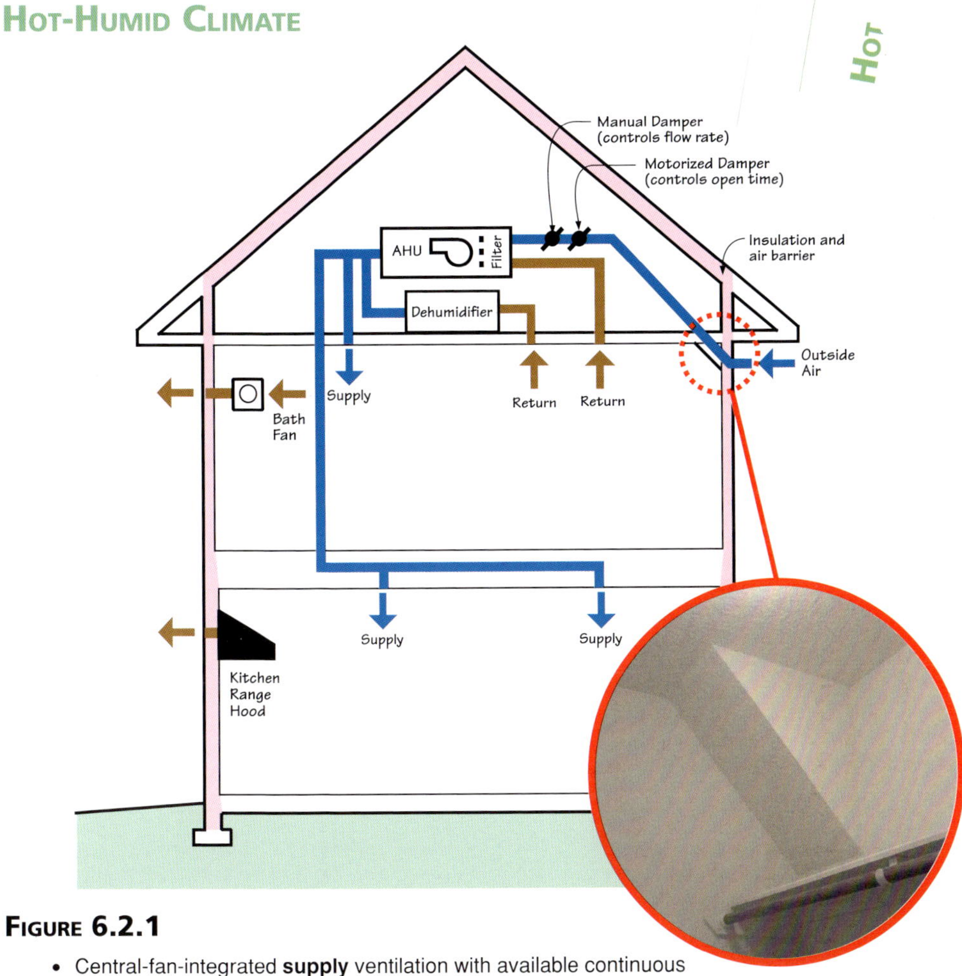

Figure 6.2.1

- Central-fan-integrated **supply** ventilation with available continuous bathroom **exhaust** and with **integrated dehumidifier**. Outside air intake through sidewall via angled fur-down in a closet (going through a gable end is just as good but avoid going through the roof).
- Manual balancing damper in outside air duct allows adjustment of the flow rate.
- Periodic operation of the central air handler fan assures consistent ventilation air distribution and uniform air quality. It also reduces temperature and humidity variations between rooms.
- Optional motorized damper closes the opening to outside when the fan is off, and with damper cycling control, can limit outside air intake independent of how long the fan runs.
- Keeping all ducts inside insulated space provides the best performance, such as permitted by the **unvented-cathedralized attic** shown above. Sealed and well insulated ducts are next best.
- Supplemental dehumidification integrated with the central air distribution system provides year-around humidity control independent of cooling system operation. A stand-alone dehumidifier can indirectly serve the whole house through use of central fan cycling.

Humid Climate

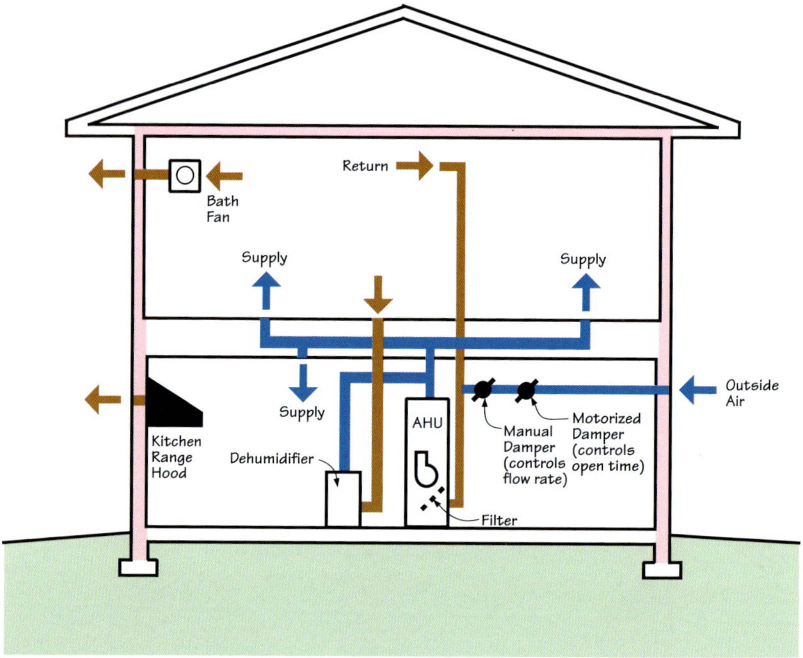

Figure 6.2.2

- Central-fan-integrated **supply** ventilation with available bathroom **exhaust** and **integrated dehumidifier**. Outside air intake through sidewall.
- Manual balancing damper in outside air duct allows adjustment of the flow rate.
- Periodic operation of the central air handler fan assures consistent ventilation air distribution and uniform air quality. It also reduces temperature and humidity variations between rooms.
- Optional motorized damper closes the opening to outside when the fan is off, and with damper cycling control, can limit outside air intake independent of how long the fan runs.
- Keeping all ducts inside insulated space provides the best performance. Sealed and well insulated ducts are next best.
- Supplemental dehumidification integrated with the central air distribution system provides year-around humidity control independent of cooling system operation.

Hot-Humid Climate

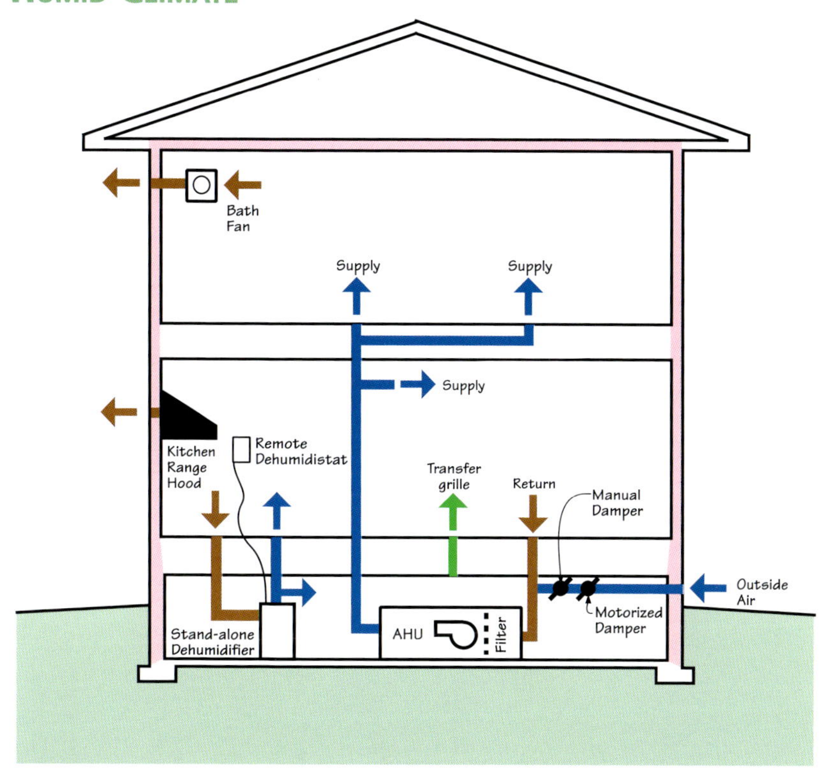

Figure 6.2.3

- Central-fan-integrated **supply** ventilation with available bathroom **exhaust** and **stand-alone dehumidifier**. Outside air intake through sidewall.
- Manual balancing damper in outside air duct allows adjustment of the flow rate.
- Periodic operation of the central air handler fan assures consistent ventilation air distribution and uniform air quality. It also reduces temperature and humidity variations between rooms.
- Optional motorized damper closes the opening to outside when the fan is off, and with damper cycling control, can limit outside air intake independent of how long the fan runs.
- Keeping all ducts inside insulated space provides the best performance, such as permitted by the sealed, **semi-conditioned crawlspace** shown above.
- Stand-alone supplemental dehumidification provides year-around humidity control independent of cooling system operation. It indirectly dehumidifies the whole house via central fan cycling. A small amount of air from the dehumidifier should be supplied to the sealed crawlspace having complete vapor barrier coverage over the floor, and poured concrete if the space will be used for storage.

Hot-Humid Climate

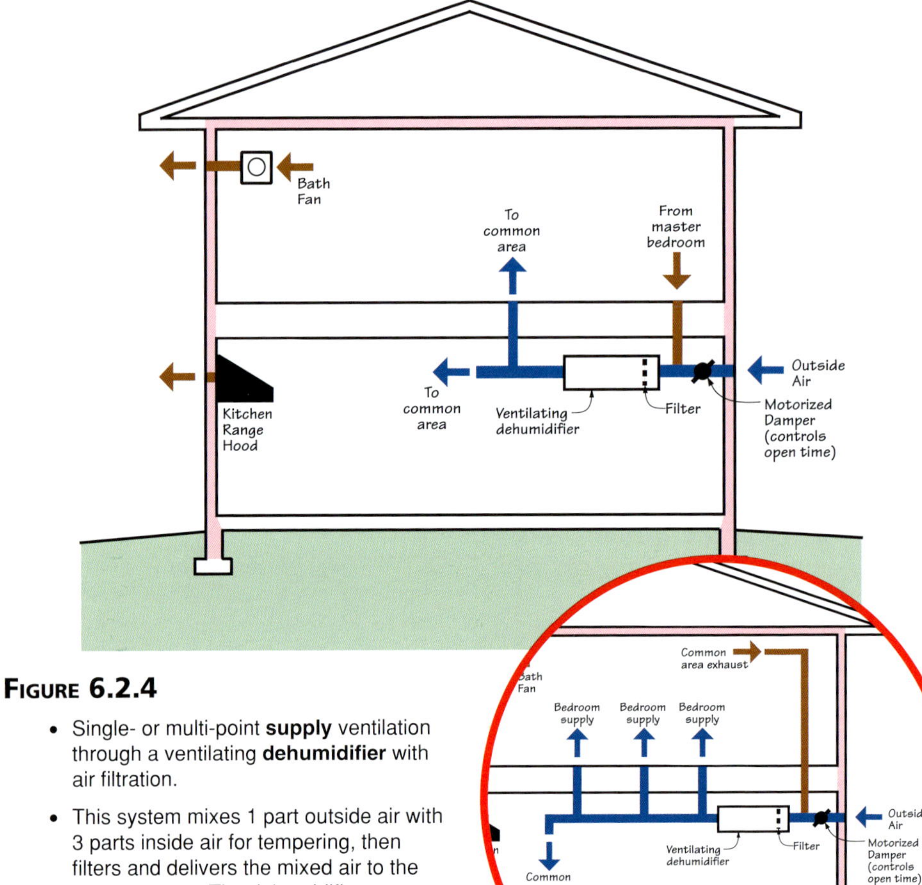

Figure 6.2.4

- Single- or multi-point **supply** ventilation through a ventilating **dehumidifier** with air filtration.
- This system mixes 1 part outside air with 3 parts inside air for tempering, then filters and delivers the mixed air to the common area. The dehumidifier compressor is activated when relative humidity in the common area rises above the dehumidistat set point.
- If single-point, the preferred recirculation air pickup should be from the master bedroom. This tends to cause ventilation air to flow back to the master bedroom through the common area. For improved air quality, periodic operation of the air handler fan provides whole-house mixing for distribution of ventilation air. It also reduces temperature variations between rooms.
- If multi-point, the recirculation air pickups should be from all bedrooms. This tends to cause ventilation air to flow back to the bedrooms through the common area. Alternately, the supply outlets could be in each room with the recirculation air pickup in the common area, however, care should be taken with supply outlets to avoid potential air discomfort over beds.
- The common area supplies should be high wall supplies to minimize any air flow discomfort.

Hot-Humid Climate

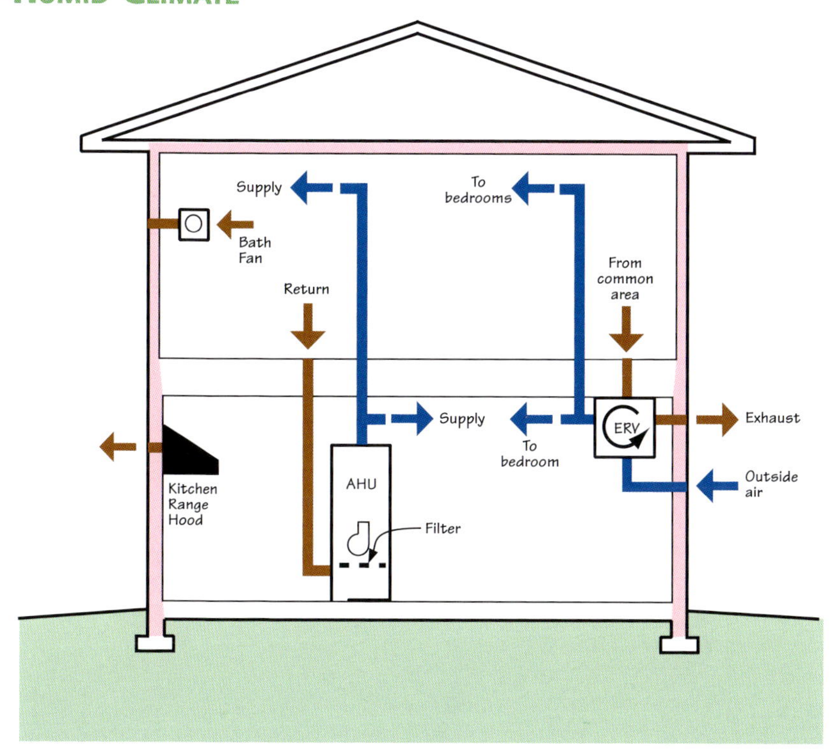

Figure 6.2.5

- Single- or multi-point **balanced** Energy Recovery Ventilator (**ERV**).
- Multi-point is best practice. The supply outlets should be in each room, avoiding air discomfort over beds, with the exhaust air pickup in the common area or bathrooms. Alternately, the exhaust air pickups could be from all bedrooms, and the ventilation air supply could be to open common areas. This tends to cause the ventilation air to flow back to the bedrooms through the common area.
- If single-point, the preferred exhaust air pickup should be from the master bedroom, and the ventilation air supply should be in the open common area. This will tend to cause the ventilation air to flow back to the master bedroom through the common area. For improved air quality, periodic operation of the air handler fan provides whole-house mixing for distribution of ventilation air. It also reduces temperature variations between rooms.
- The outside air intake and exhaust outlet should be separated by about 10 feet or use manufacturer's recommendation to avoid contamination of ventilation air.

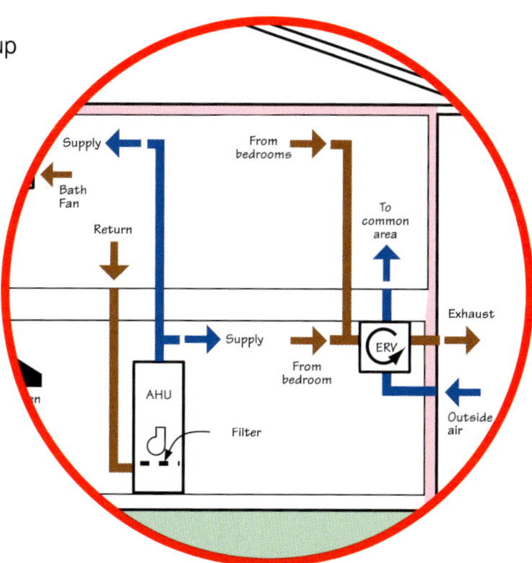

Hot-Humid Climate

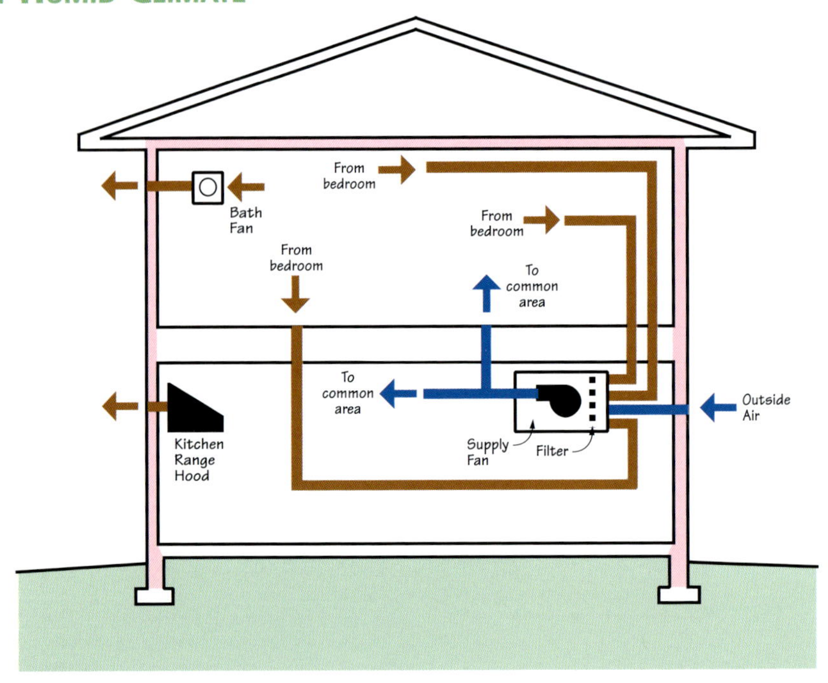

Figure 6.2.6

- Single- or multi-point **supply** ventilation system using a remote-mounted packaged fan with filter.
- This system mixes 1 part outside air with 3 parts inside air for tempering, then filters and delivers the mixed air to the common area.
- If single-point, the preferred recirculation air pickup should be from the master bedroom. This tends to cause ventilation air to flow back to the master bedroom through the common area. For improved air quality, periodic operation of the air handler fan provides whole-house mixing for distribution of ventilation air. It also reduces temperature variations between rooms.
- If multi-point, the recirculation air pickups should be from all bedrooms. This tends to cause ventilation air to flow back to the bedrooms through the common area. Alternately, the supply outlets could be in each room with the recirculation air pickup in the common area, however, care should be taken with supply outlets to avoid potential air discomfort over beds.
- The common area supplies should be high wall supplies to minimize any air flow discomfort.

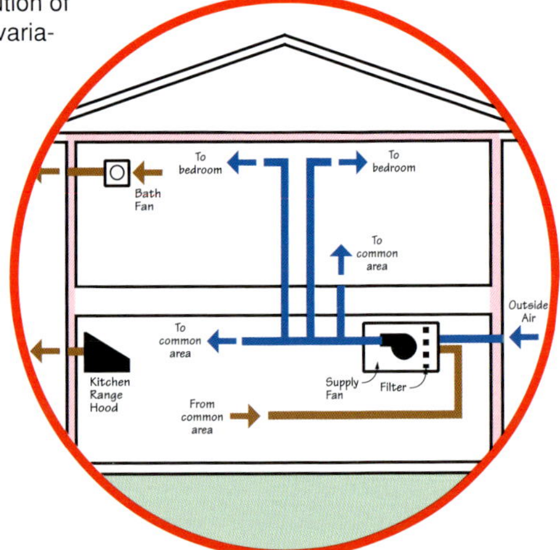

Hot-Humid Climate

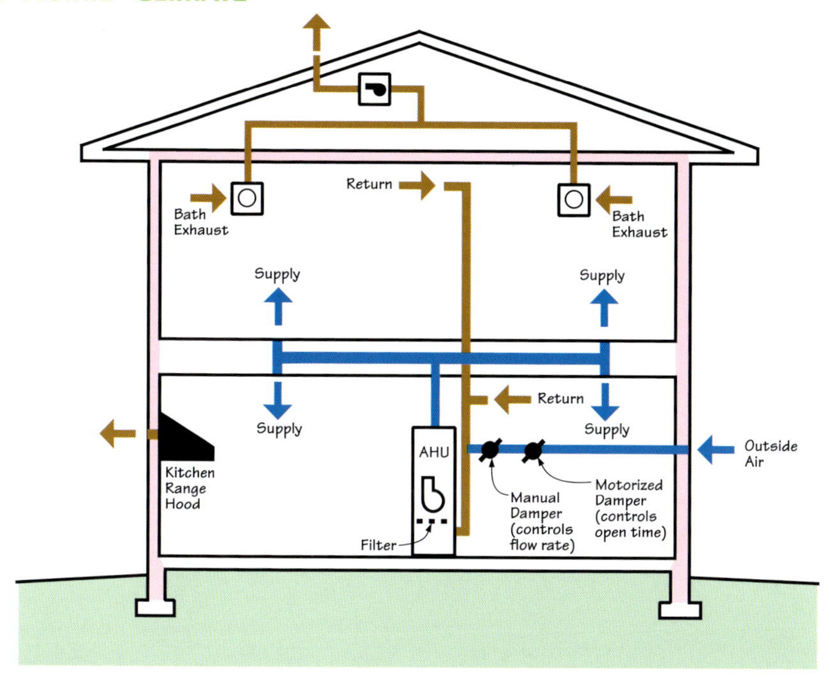

Figure 6.2.7

- Central-fan-integrated **supply** ventilation with multi-point **exhaust** from bathrooms.
- Occupant activated bathroom switches change the fan to high speed for a time.
- Manual balancing damper in outside air duct allows adjustment of the flow rate.
- Periodic operation of the central air handler fan assures consistent ventilation air distribution and uniform air quality. It also reduces temperature variations between rooms.
- Optional motorized damper closes the opening to outside when the fan is off. With damper cycling control, outside air intake can be limited independently of how long the fan runs.
- Keeping all ducts inside insulated space provides the best performance. Sealed and well insulated ducts are next best.

Hot-Humid Climate

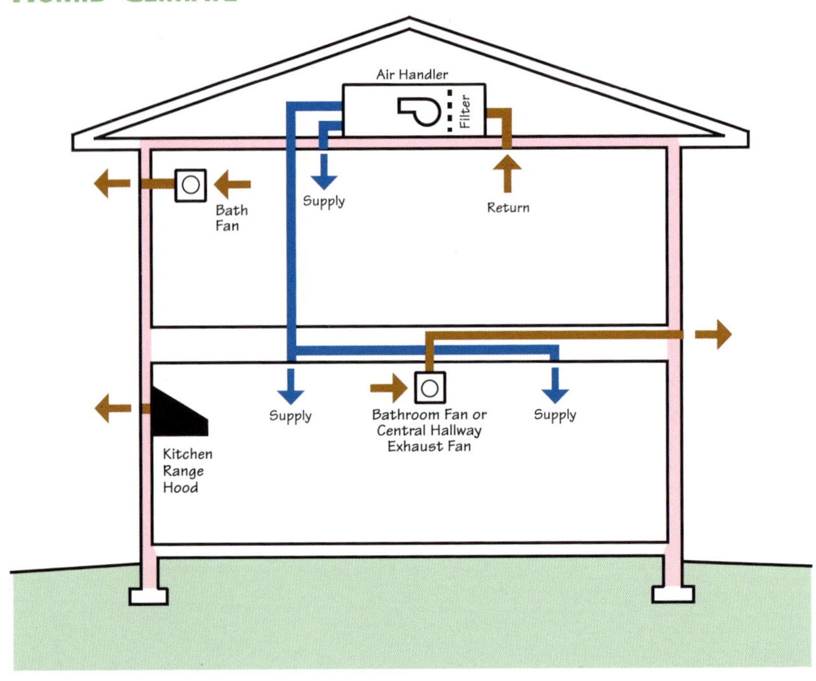

Figure 6.2.8

- Single-point **exhaust** ventilation, located preferably in a central hallway or common area. Preferably not located in a utility room off the garage.
- For improved air quality, periodic operation of the air handler fan provides whole-house mixing for distribution of ventilation air. It also reduces temperature variations between rooms.
- Continuous exhaust is discouraged in favor of supply or balanced ventilation in cooled buildings in hot-humid climates. Drawing high humidity air inward through the building enclosure, where it may contact cool surfaces and condense, can lead to moisture and indoor air quality problems. Continuous exhaust with powerful fans should not be used.

Mixed Climate

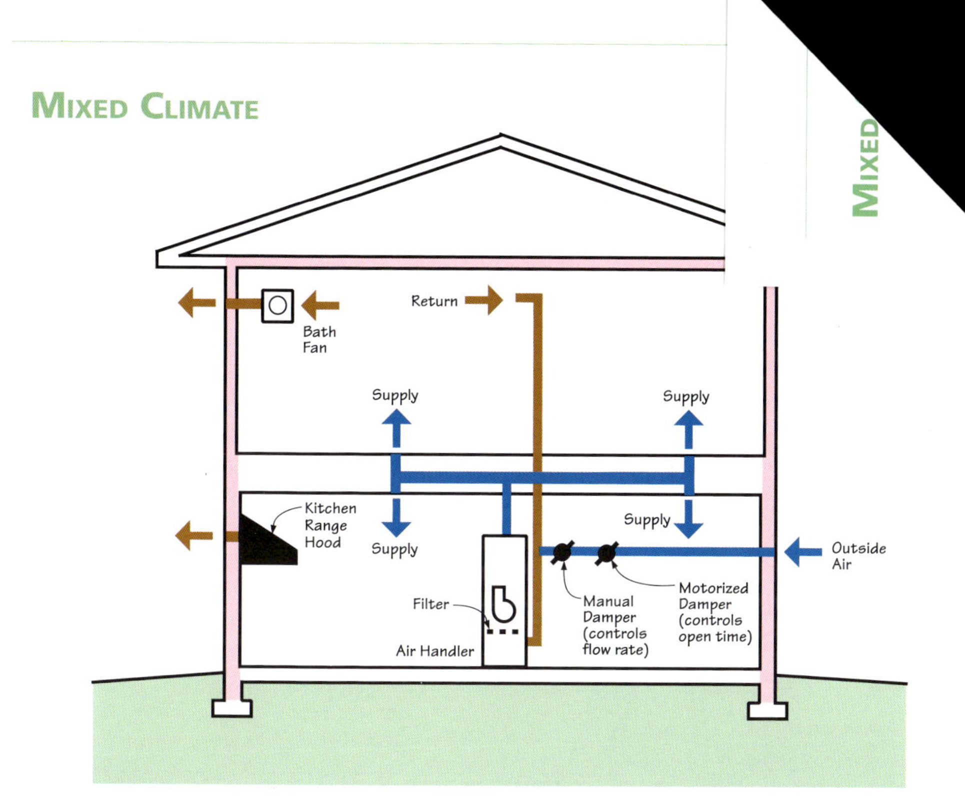

Figure 6.3.1

- Central-fan-integrated **supply** ventilation with available bathroom **exhaust**. Outside air intake through sidewall.
- Manual balancing damper in outside air duct allows adjustment of the flow rate.
- Periodic operation of the central air handler fan assures consistent ventilation air distribution and uniform air quality. It also reduces temperature variations between rooms.
- Optional motorized damper closes the opening to outside when the fan is off. With damper cycling control, outside air intake can be limited independently of how long the fan runs.
- Keeping all ducts inside insulated space provides the best performance. Sealed and well insulated ducts are next best.

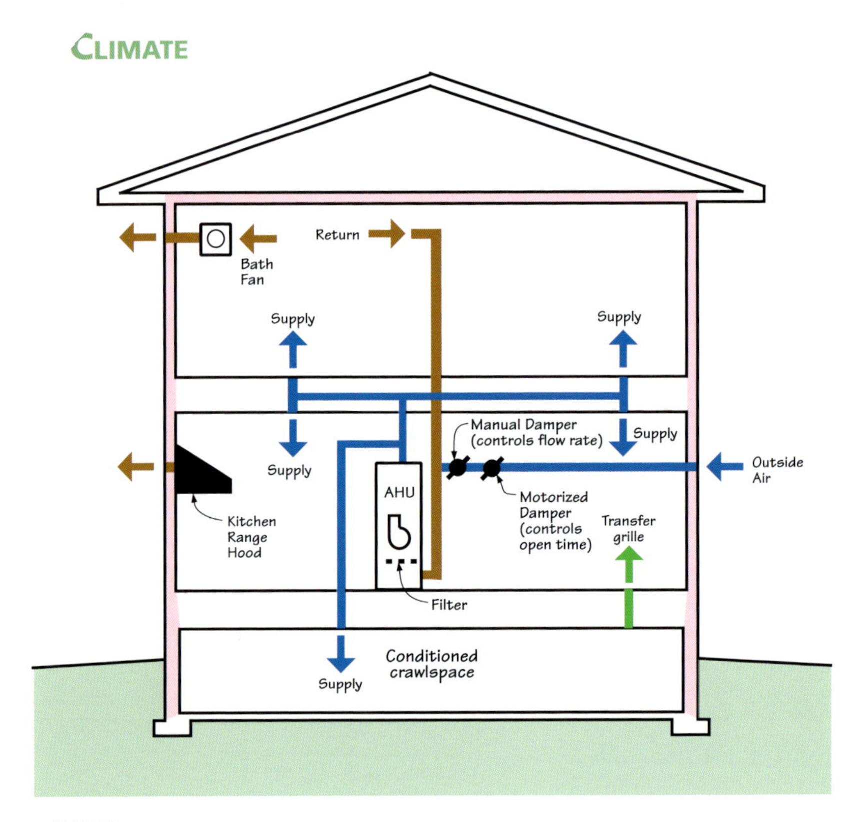

FIGURE 6.3.2

- Central-fan-integrated **supply** ventilation with available bathroom **exhaust**. Outside air intake through sidewall.
- Manual balancing damper in outside air duct allows adjustment of the flow rate.
- Periodic operation of the central air handler fan assures consistent ventilation air distribution and uniform air quality. It also reduces temperature variations between rooms.
- Optional motorized damper closes the opening to outside when the fan is off. With damper cycling control, outside air intake can be limited independently of how long the fan runs.
- Keeping all ducts inside insulated space provides the best performance, such as permitted by the sealed, **semi-conditioned crawlspace** shown above.
- A small amount of conditioned air should be supplied to the sealed crawlspace having complete vapor barrier coverage over the floor, and poured concrete if the space will be used for storage.

Mixed Climate

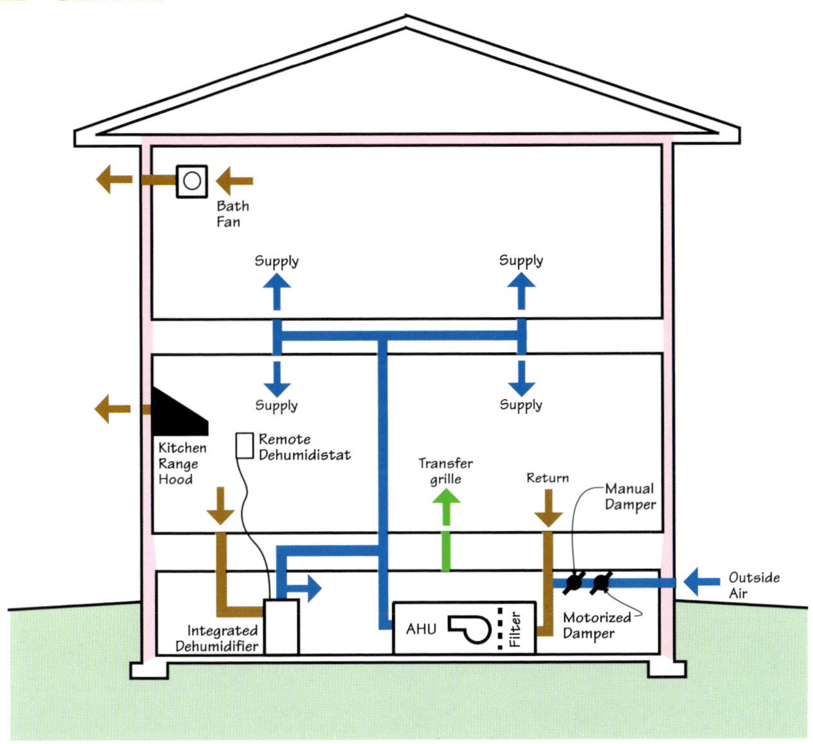

Figure 6.3.3

- Central-fan-integrated **supply** ventilation with available bathroom **exhaust** and with **integrated dehumidifier**. Outside air intake through sidewall.
- Manual balancing damper in outside air duct allows adjustment of the flow rate.
- Periodic operation of the central air handler fan assures consistent ventilation air distribution and uniform air quality. It also reduces temperature variations between rooms.
- Optional motorized damper closes the opening to outside when the fan is off. With damper cycling control, outside air intake can be limited independently of how long the fan runs.
- Keeping all ducts inside insulated space provides the best performance, such as permitted by the sealed, **semi-conditioned crawlspace** shown above.
- Supplemental dehumidification integrated with the central air distribution system provides year-around humidity control independent of cooling system operation. A small amount of air from the dehumidifier should be supplied to the sealed crawlspace having complete vapor barrier coverage over the floor, and poured concrete if the space will be used for storage.

Mixed Climate

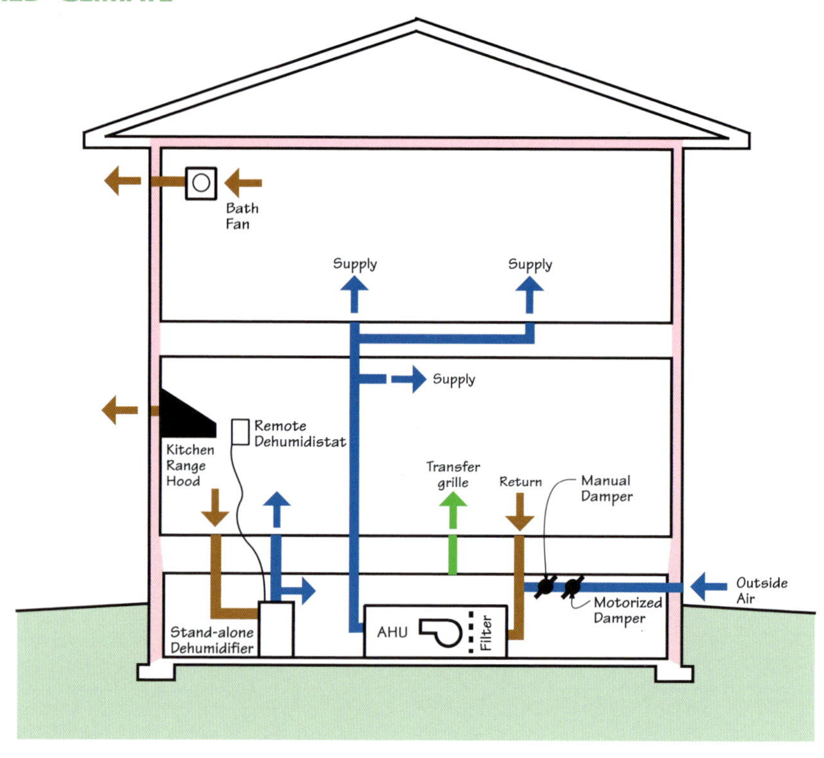

Figure 6.3.4

- Central-fan-integrated **supply** ventilation with available bathroom **exhaust** and with **stand-alone dehumidifier**. Outside air intake through sidewall.
- Manual balancing damper in outside air duct allows adjustment of the flow rate.
- Periodic operation of the central air handler fan assures consistent ventilation air distribution and uniform air quality. It also reduces temperature variations between rooms.
- Optional motorized damper closes the opening to outside when the fan is off. With damper cycling control, outside air intake can be limited independently of how long the fan runs.
- Keeping all ducts inside insulated space provides the best performance, such as permitted by the sealed, **semi-conditioned crawlspace** shown above.
- Stand-alone supplemental dehumidification provides year-around humidity control independent of cooling system operation. It indirectly dehumidifies the whole house via central fan cycling. A small amount of air from the dehumidifier should be supplied to the sealed crawlspace having complete vapor barrier coverage over the floor, and poured concrete if the space will be used for storage.

Mixed Climate

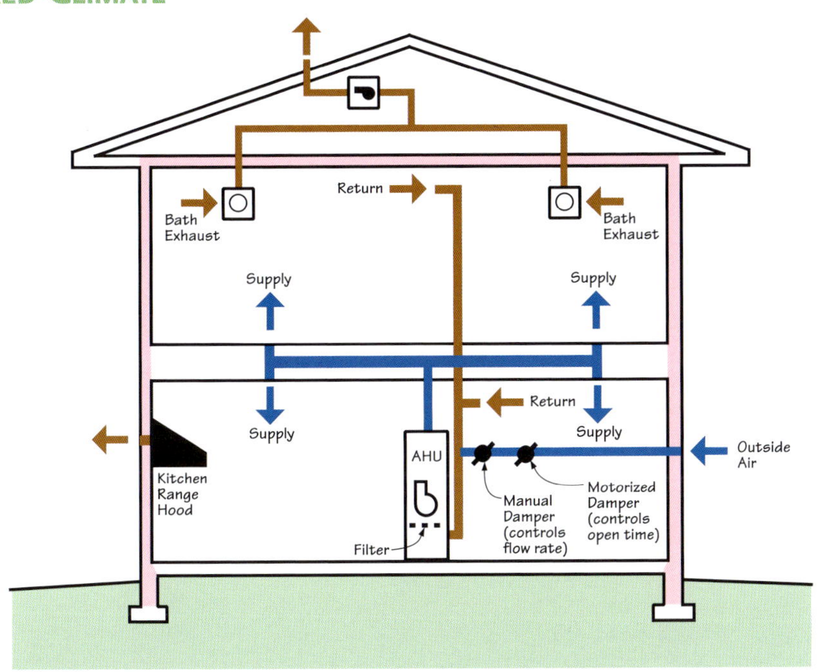

Figure 6.3.5

- Central-fan-integrated **supply** ventilation with multi-point **exhaust** from bathrooms.
- Occupant activated bathroom switches change the fan to high speed for a time.
- Manual balancing damper in outside air duct allows adjustment of the flow rate.
- Periodic operation of the central air handler fan assures consistent ventilation air distribution and uniform air quality. It also reduces temperature variations between rooms.
- Optional motorized damper closes the opening to outside when the fan is off. With damper cycling control, outside air intake can be limited independently of how long the fan runs.
- Keeping all ducts inside insulated space provides the best performance. Sealed and well insulated ducts are next best.

Mixed Climate

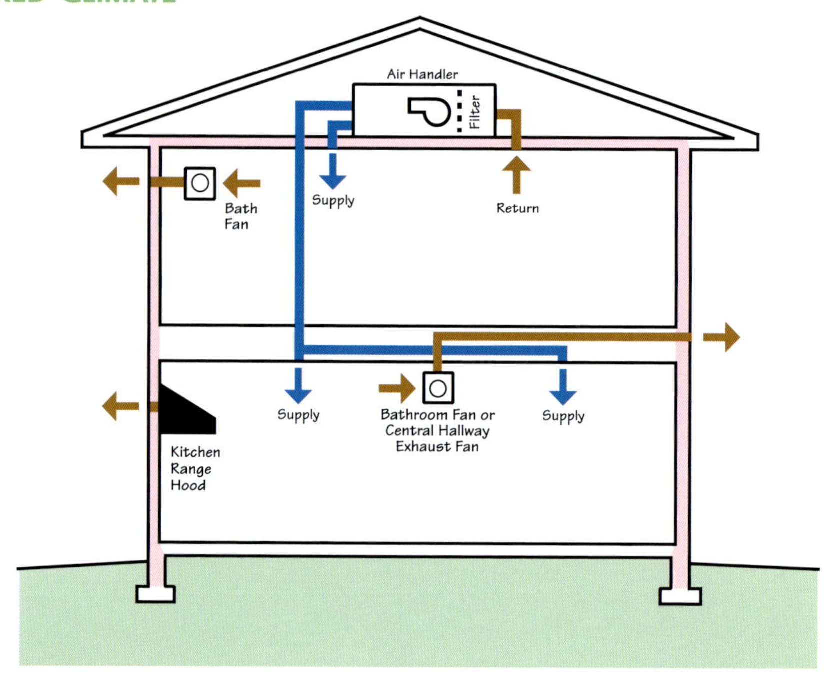

Figure 6.3.6

- Single-point **exhaust** ventilation, located preferably in a central hallway or common area. Preferably not located in a utility room off the garage.
- For improved air quality, periodic operation of the air handler fan provides whole-house mixing for distribution of ventilation air. It also reduces temperature variations between rooms.

Mixed Climate

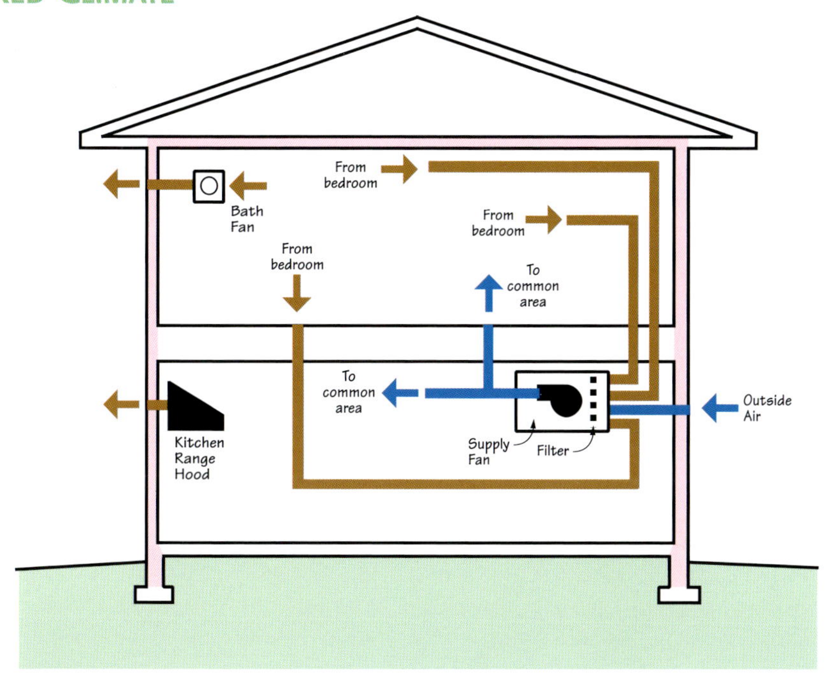

Figure 6.3.7

- Single- or multi-point **supply** ventilation system using a remote-mounted packaged fan with filter.
- This system mixes 1 part outside air with 3 parts inside air for tempering, then filters and delivers the mixed air to the common area.
- If single-point, the preferred recirculation air pickup should be from the master bedroom. This tends to cause ventilation air to flow back to the master bedroom through the common area. For improved air quality, periodic operation of the air handler fan provides whole-house mixing for distribution of ventilation air. It also reduces temperature variations between rooms.
- If multi-point, the recirculation air pickups should be from all bedrooms. This tends to cause ventilation air to flow back to the bedrooms through the common area. Alternately, the supply outlets could be in each room with the recirculation air pickup in the common area, however, care should be taken with supply outlets to avoid potential air discomfort over beds.
- The common area supplies should be high wall supplies to minimize any air flow discomfort.

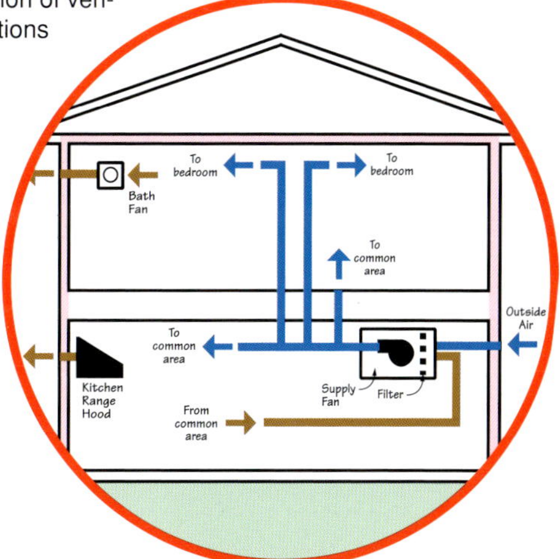

Mixed Climate

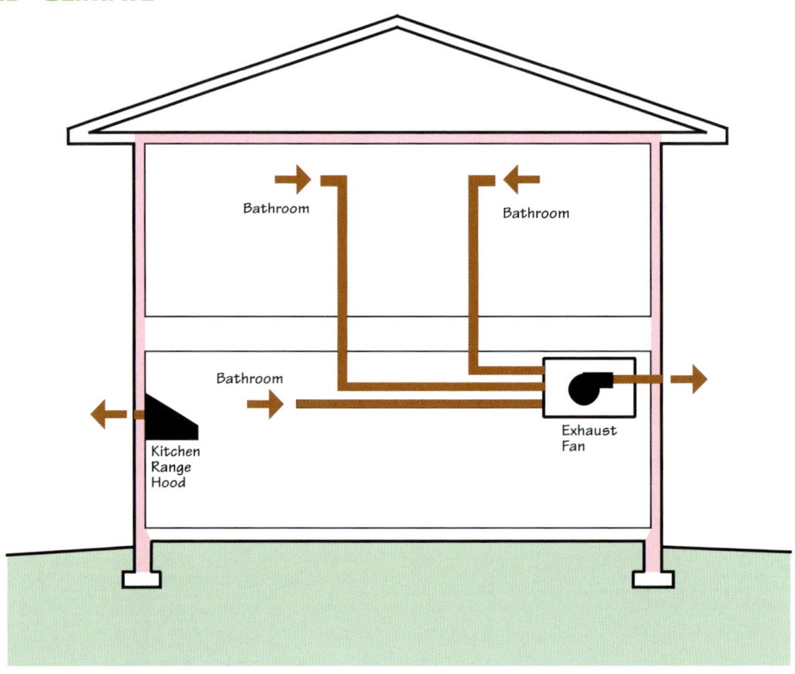

Figure 6.3.8

- Multi-point **exhaust** pulling air continuously from each bathroom.
- Occupant activated bathroom switches change the fan to high speed for a time.
- Best used in mixed-dry climates rather than mixed-humid. Continuous exhaust that depressurizes the building is discouraged in favor of supply or balanced ventilation in cooled buildings in mixed-humid climates. Drawing high humidity air inward through the building enclosure, where it may contact cool surfaces and condense, can lead to moisture and indoor air quality problems.
- Best practice is to not locate natural draft combustion appliances in the indoor environment. As with all exhaust only ventilation systems, special attention should be given in order to prevent problems such as backdrafting of combustion appliances, intaking of soil gases, or intaking of air from an attached garage. The larger the exhaust air flow, the more the concern.

Cold Climate

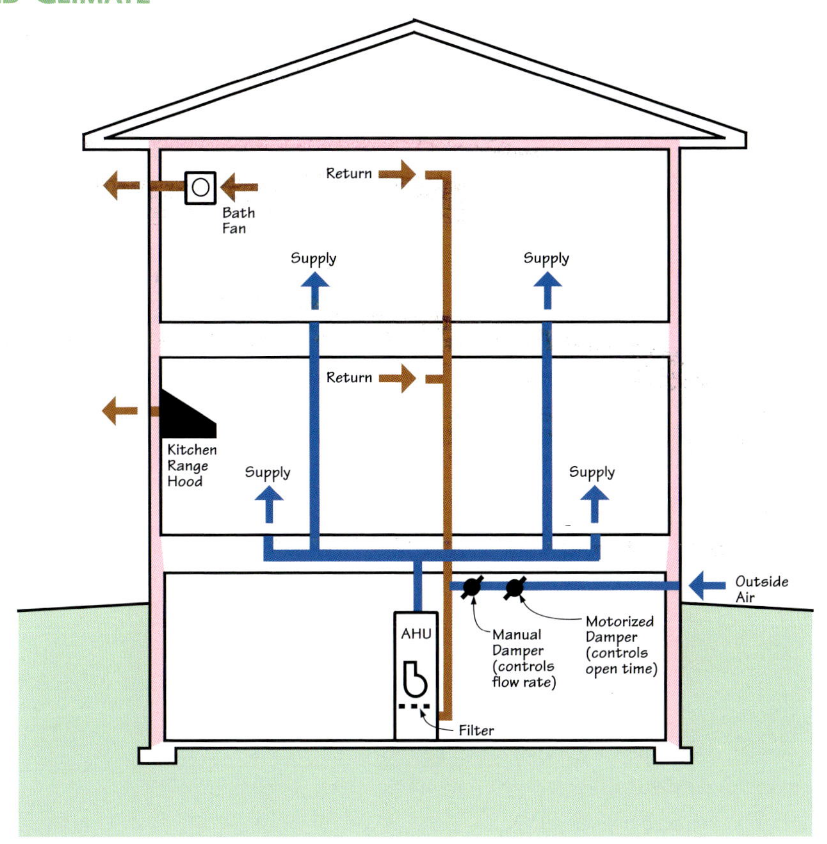

Figure 6.4.1

- Central-fan-integrated **supply** ventilation with available bathroom **exhaust**.
- Manual balancing damper in outside air duct allows adjustment of the flow rate. Limit the outside air fraction to about 15% (see **Figure 1** in Section 3.1.1).
- Periodic operation of the central air handler fan assures consistent ventilation air distribution and uniform air quality. It also reduces temperature variations between rooms.
- Optional motorized damper closes the opening to outside when the fan is off. With damper cycling control, outside air intake can be limited independently of how long the fan runs.
- Keeping all ducts inside insulated space provides the best performance. Sealed and well insulated ducts are next best.

Cold Climate

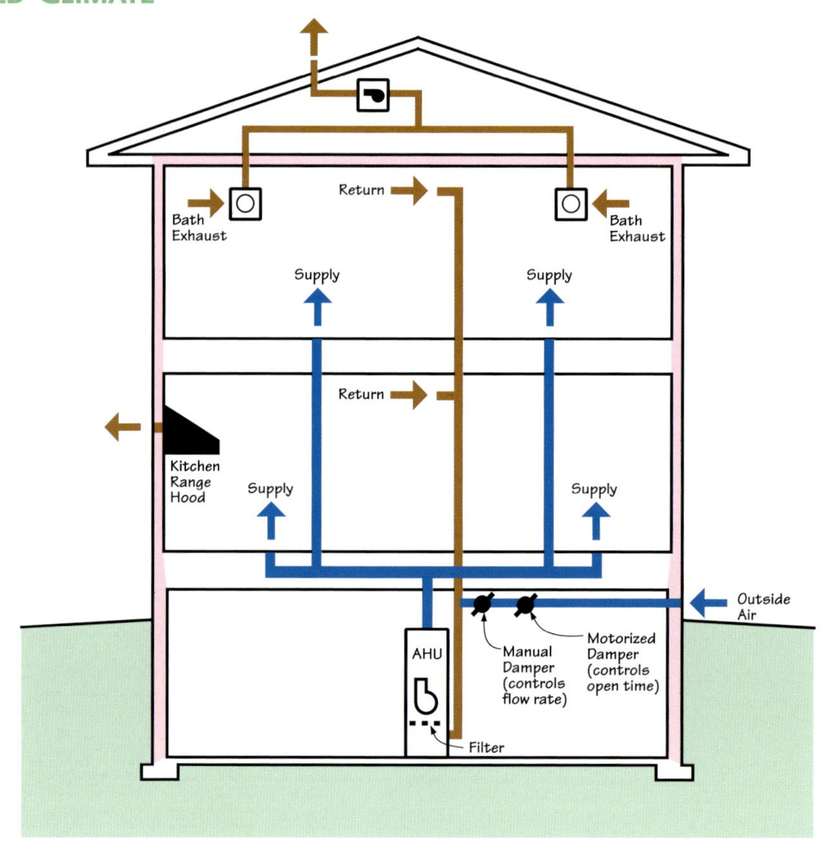

Figure 6.4.2

- Central-fan-integrated **supply** ventilation with multi-point **exhaust** from bathrooms.
- Occupant activated bathroom switches change the fan to high speed for a time.
- Manual balancing damper in outside air duct allows adjustment of the flow rate. Limit the outside air fraction to about 15% (see **Figure 1** in Section 3.1.1).
- Periodic operation of the central air handler fan assures consistent ventilation air distribution and uniform air quality. It also reduces temperature variations between rooms.
- Optional motorized damper closes the opening to outside when the fan is off. With damper cycling control, outside air intake can be limited independently of how long the fan runs.
- Keeping all ducts inside insulated space provides the best performance. Sealed and well insulated ducts are next best.

Cold Climate

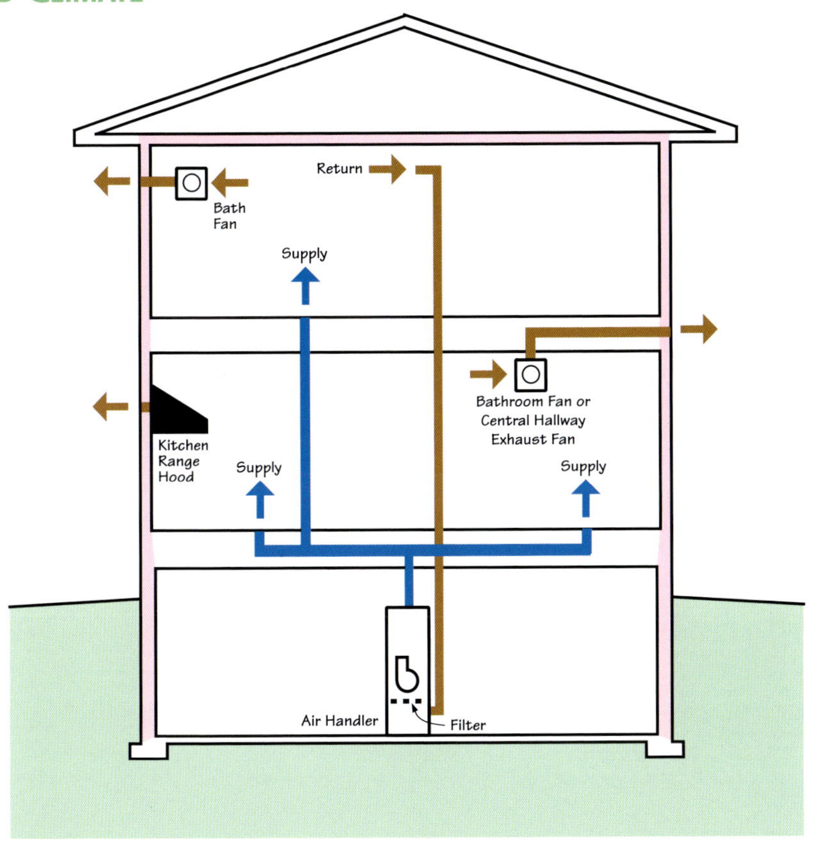

Figure 6.4.3

- Single-point **exhaust** ventilation, located preferably in a central hallway or common area. Preferably not located in a utility room off the garage.
- For improved air quality, periodic operation of the air handler fan provides whole-house mixing for distribution of ventilation air. It also reduces temperature variations between rooms.

Cold Climate

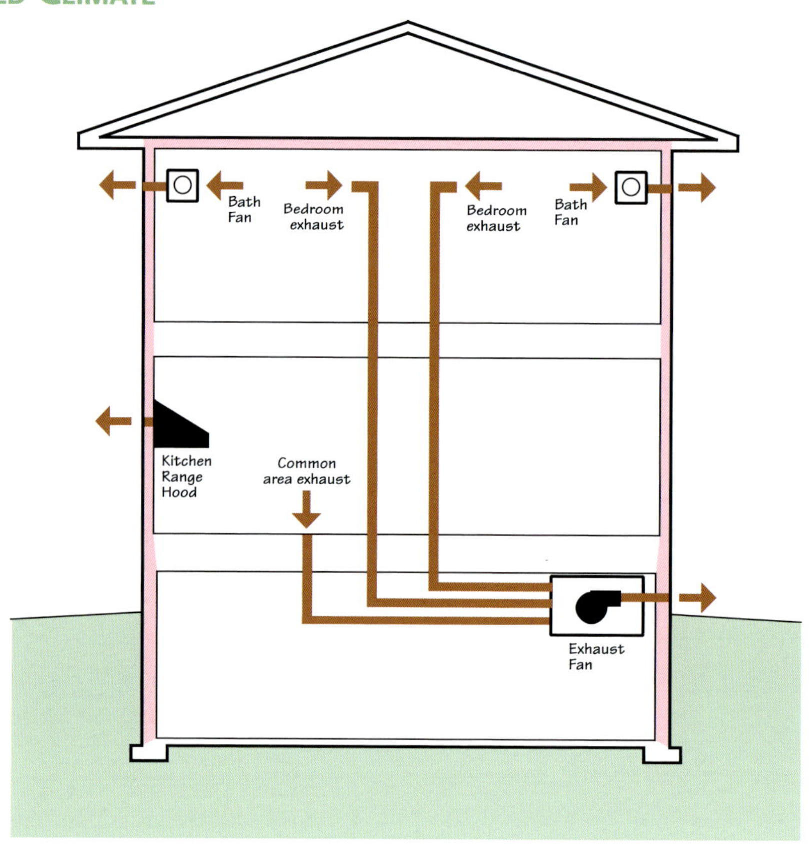

Figure 6.4.4

- Multi-point **exhaust** pulling air continuously from each bedroom and the common area.
- Best practice is to not locate natural draft combustion appliances in the indoor environment. As with all exhaust only ventilation systems, special attention should be given in order to prevent problems such as backdrafting of combustion appliances, intaking of soil gases, or intaking of air from an attached garage. The larger the exhaust air flow, the more the concern.

Cold Climate

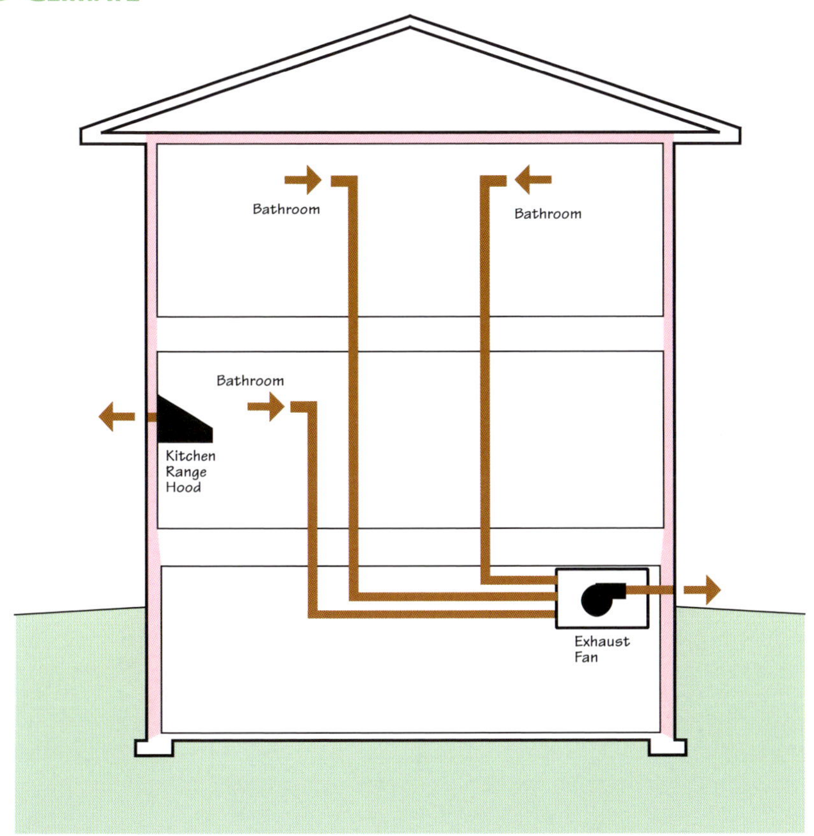

Figure 6.4.5

- Multi-point **exhaust** pulling air continuously from each bathroom.
- A switch in the bathroom changes the fan to high speed for a time.
- Best practice is to not locate natural draft combustion appliances in the indoor environment. As with all exhaust only ventilation systems, special attention should be given in order to prevent problems such as backdrafting of combustion appliances, intaking of soil gases, or intaking of air from an attached garage. The larger the exhaust air flow, the more the concern.

Cold Climate

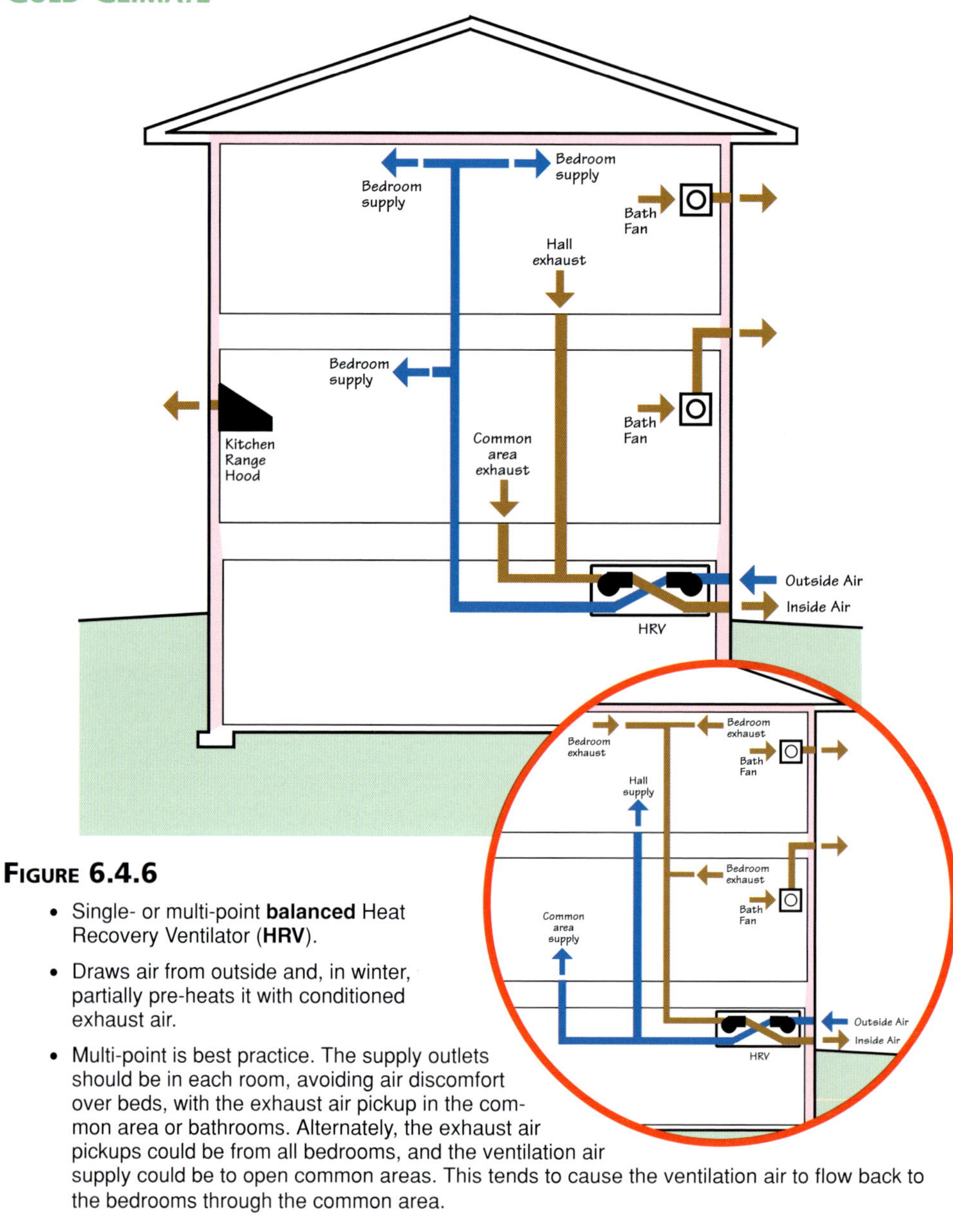

Figure 6.4.6

- Single- or multi-point **balanced** Heat Recovery Ventilator (**HRV**).
- Draws air from outside and, in winter, partially pre-heats it with conditioned exhaust air.
- Multi-point is best practice. The supply outlets should be in each room, avoiding air discomfort over beds, with the exhaust air pickup in the common area or bathrooms. Alternately, the exhaust air pickups could be from all bedrooms, and the ventilation air supply could be to open common areas. This tends to cause the ventilation air to flow back to the bedrooms through the common area.
- If single-point, the preferred exhaust air pickup should be from the master bedroom, and the ventilation air supply should be in the open common area. This will tend to cause the ventilation air to flow back to the master bedroom through the common area. For improved air quality, periodic operation of the air handler fan provides whole-house mixing for distribution of ventilation air. It also reduces temperature variations between rooms.
- The outside air intake and exhaust outlet should be separated by about 10 feet or use manufacturer's recommendation to avoid contamination of ventilation air.

Cold Climate

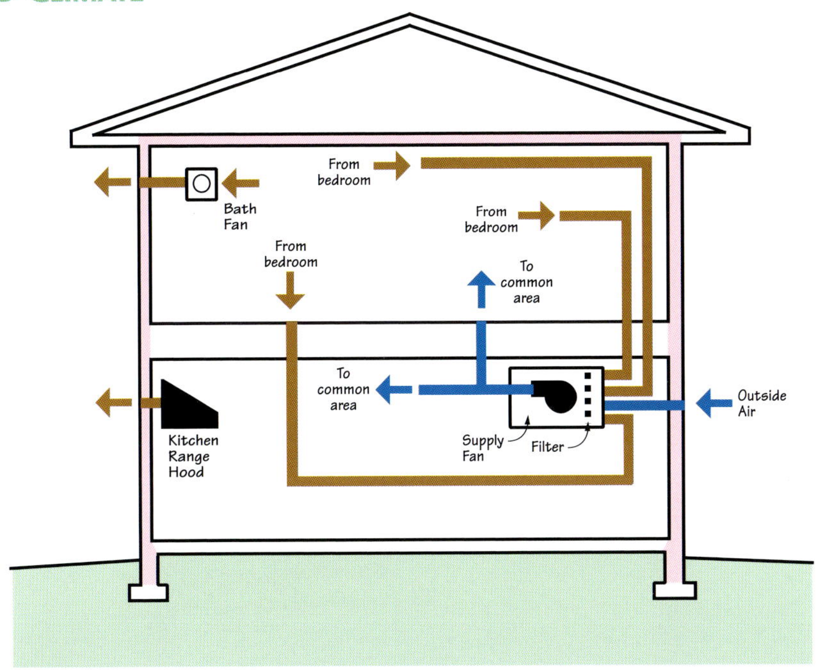

Figure 6.4.7

- Single- or multi-point **supply** ventilation system using a remote-mounted packaged fan with filter.
- This system mixes 1 part outside air with 3 parts inside air for tempering, then filters and delivers the mixed air to the common area.
- If single-point, the preferred recirculation air pickup should be from the master bedroom. This tends to cause ventilation air to flow back to the master bedroom through the common area. For improved air quality, periodic operation of the air handler fan provides whole-house mixing for distribution of ventilation air. It also reduces temperature variations between rooms.
- If multi-point, the recirculation air pickups should be from all bedrooms. This tends to cause ventilation air to flow back to the bedrooms through the common area. Alternately, the supply outlets could be in each room with the recirculation air pickup in the common area, however, care should be taken with supply outlets to avoid potential air discomfort over beds.
- The common area supplies should be high wall supplies to minimize any air flow discomfort.

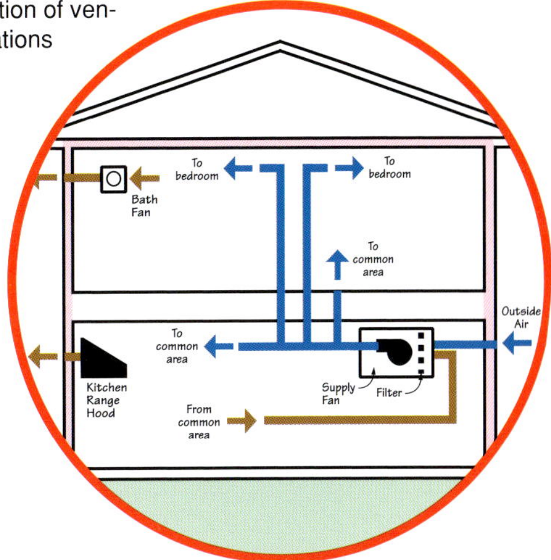

Very Cold Climate

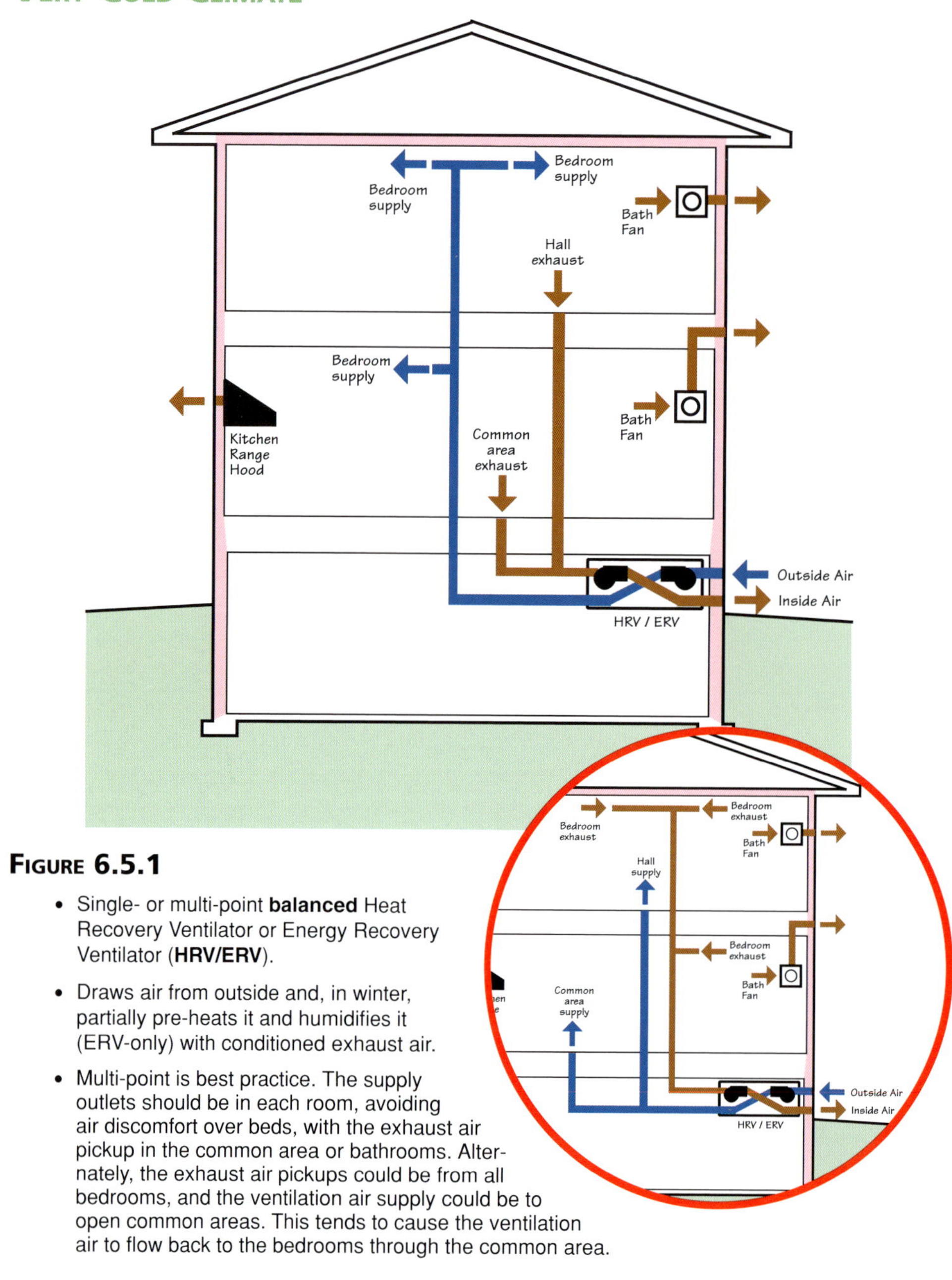

Figure 6.5.1

- Single- or multi-point **balanced** Heat Recovery Ventilator or Energy Recovery Ventilator (**HRV/ERV**).
- Draws air from outside and, in winter, partially pre-heats it and humidifies it (ERV-only) with conditioned exhaust air.
- Multi-point is best practice. The supply outlets should be in each room, avoiding air discomfort over beds, with the exhaust air pickup in the common area or bathrooms. Alternately, the exhaust air pickups could be from all bedrooms, and the ventilation air supply could be to open common areas. This tends to cause the ventilation air to flow back to the bedrooms through the common area.

Very Cold Climate

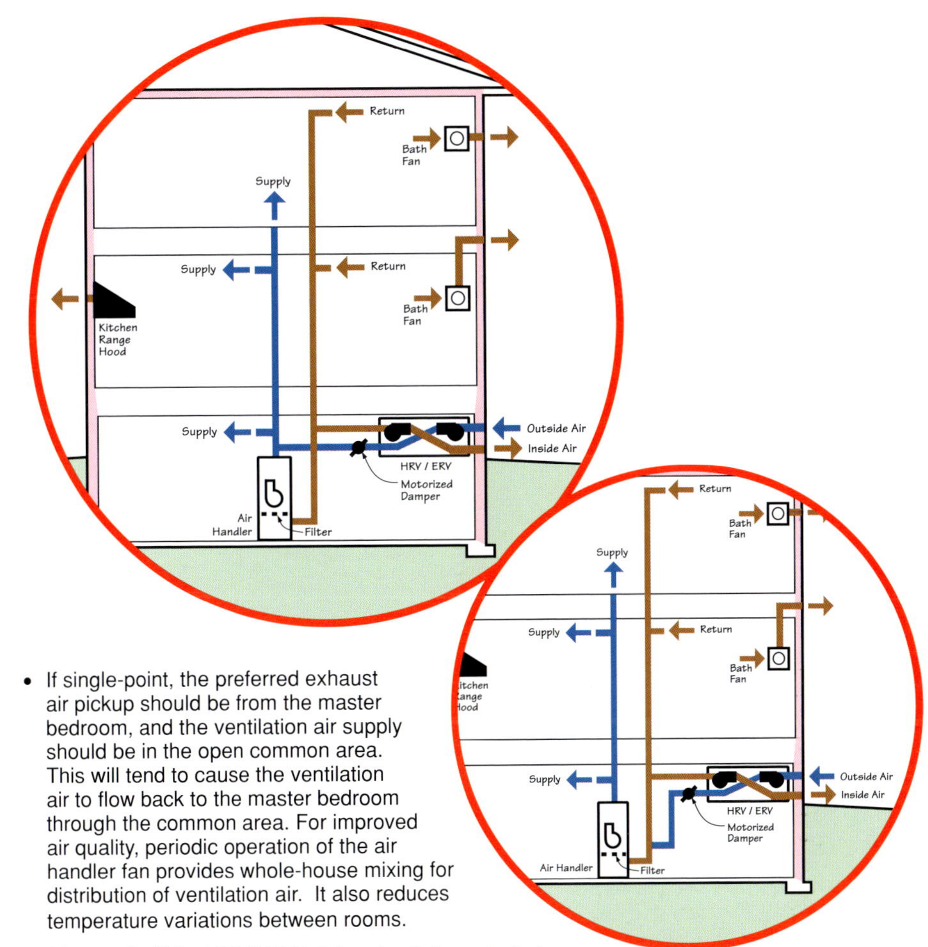

- If single-point, the preferred exhaust air pickup should be from the master bedroom, and the ventilation air supply should be in the open common area. This will tend to cause the ventilation air to flow back to the master bedroom through the common area. For improved air quality, periodic operation of the air handler fan provides whole-house mixing for distribution of ventilation air. It also reduces temperature variations between rooms.
- Alternately, if the HRV/ERV inlet and outlet are ducted to the central return and supply ducts, then coincident operation of the central fan is required with operation of the HRV/ERV. For this reason an independent, fully ducted HRV/ERV system is more efficient.
- Alternately, for cold and very cold climates, the HRV/ERV exhaust air pick ups may be from the bathrooms, and the HRV/ERV outlet may be into either the furnace return or supply.

Very Cold Climate

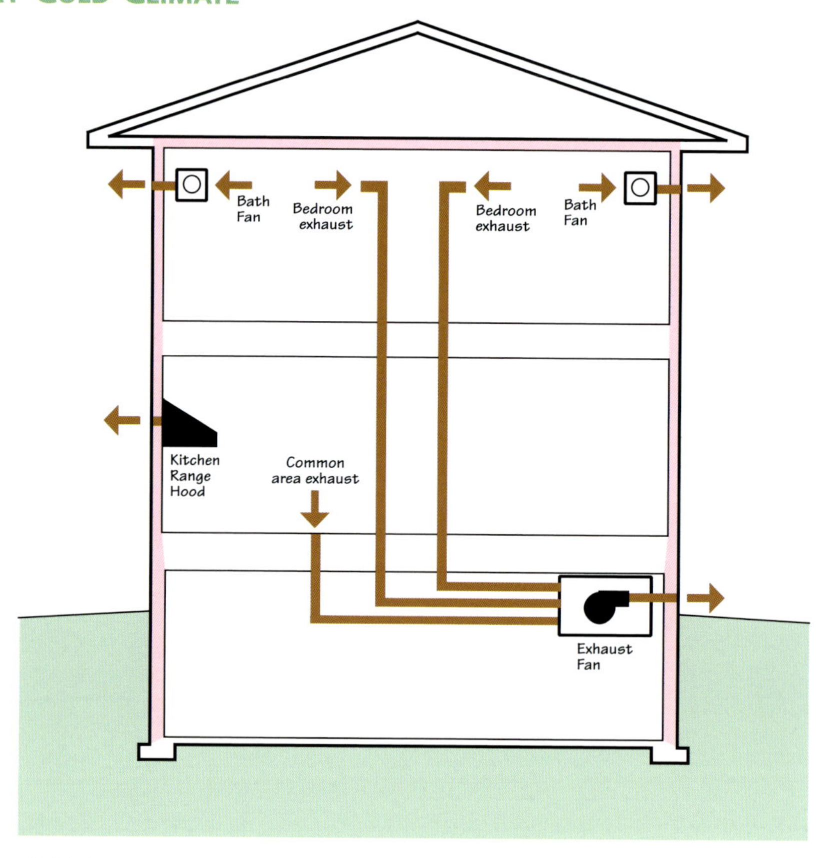

Figure 6.5.2

- Multi-point **exhaust** pulling air continuously from each bedroom and the common area.
- Best practice is to not locate natural draft combustion appliances in the indoor environment. As with all exhaust only ventilation systems, special attention should be given in order to prevent problems such as backdrafting of combustion appliances, intaking of soil gases, or intaking of air from an attached garage. The larger the exhaust air flow, the more the concern.

Very Cold Climate

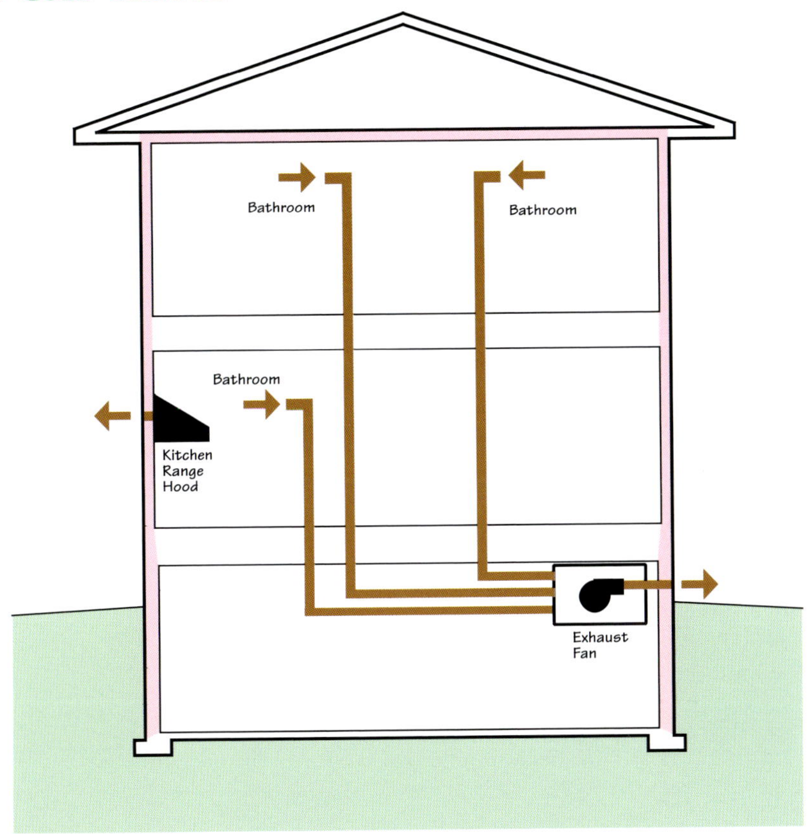

Figure 6.5.3

- Multi-point **exhaust** pulling air continuously from each bathroom.
- A switch in the bathroom changes the fan to high speed for a time.
- Best practice is to not locate natural draft combustion appliances in the indoor environment. As with all exhaust only ventilation systems, special attention should be given in order to prevent problems such as backdrafting of combustion appliances, intaking of soil gases, or intaking of air from an attached garage. The larger the exhaust air flow, the more the concern.

Very Cold Climate

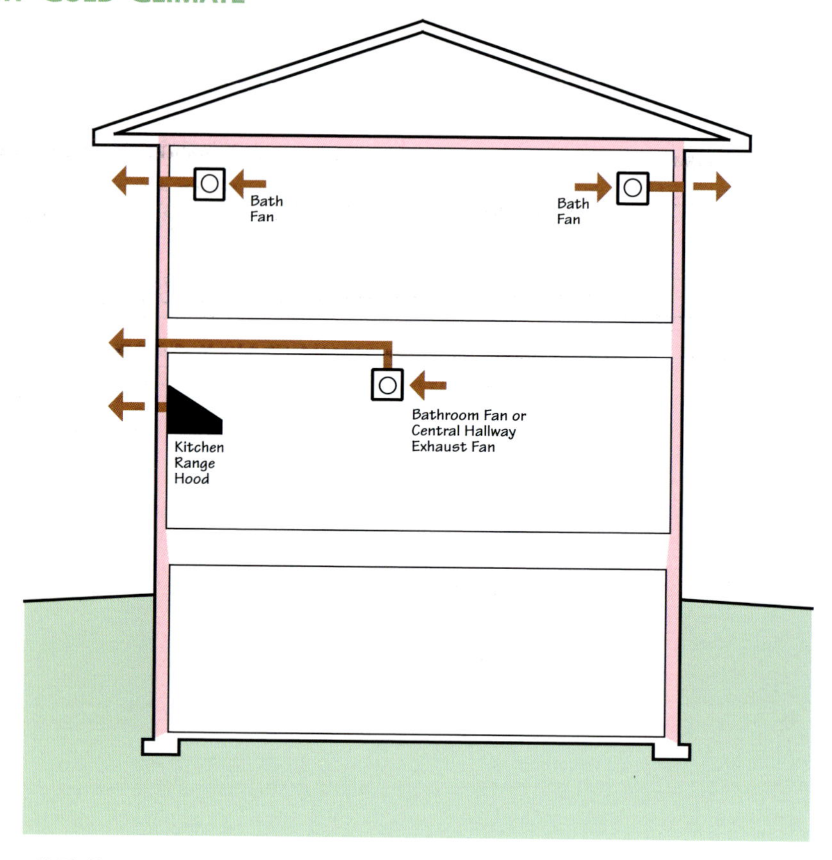

Figure 6.5.4

- Single-point **exhaust** ventilation, located preferably in a central hallway or common area. Preferably not located in a utility room off the garage.
- If a furnace air distribution system is present, then for improved air quality, periodic operation of the air handler fan provides whole-house mixing for distribution of ventilation air. It also reduces temperature variations between rooms.

Very Cold Climate

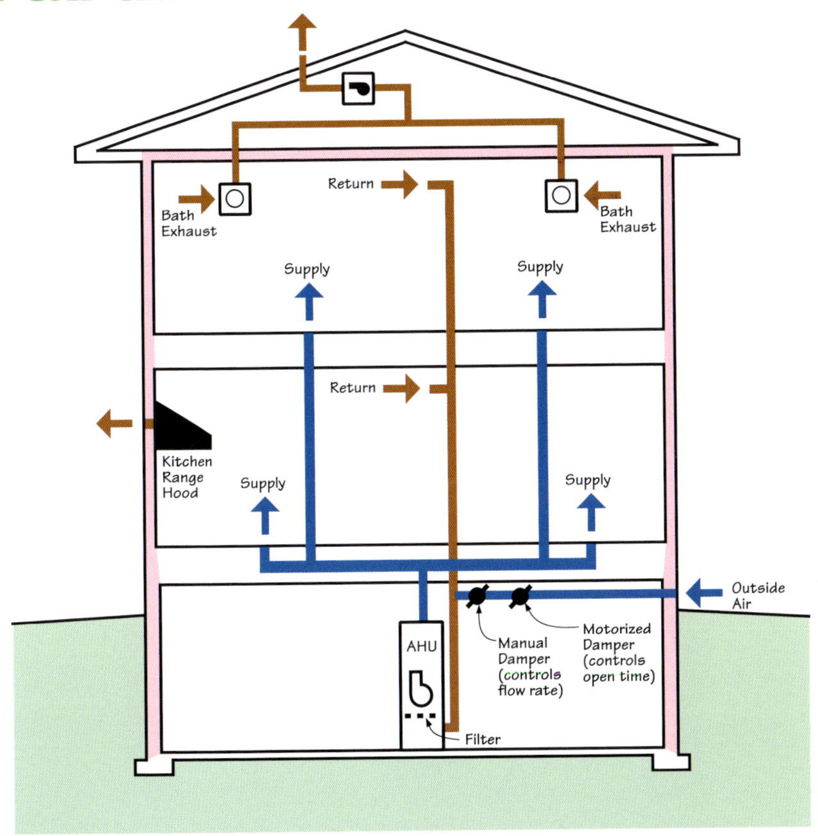

Figure 6.5.5

- Central-fan-integrated **supply** ventilation with multi-point **exhaust** from bathrooms.
- Occupant activated bathroom switches change the fan to high speed for a time.
- Manual balancing damper in outside air duct allows adjustment of the flow rate. Limit the outside air fraction to about 7% of total air handler flow (see **Figure 1** in Section 3.1.1).
- Periodic operation of the central air handler fan assures consistent ventilation air distribution and uniform air quality. It also reduces temperature variations between rooms.
- A motorized damper is required to close the opening to outside when the fan is off. With damper cycling control, outside air intake can be limited independently of how long the fan runs.
- Keeping all ducts inside insulated space provides the best performance. Sealed and well insulated ducts are next best.

Very Cold Climate

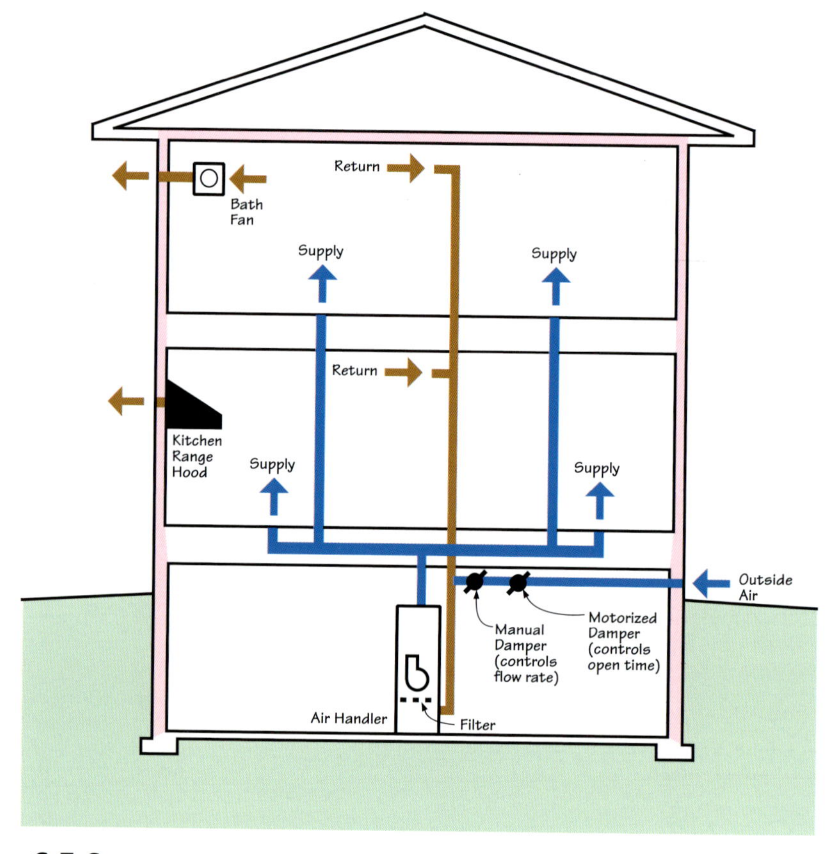

Figure 6.5.6

- Central-fan-integrated **supply** ventilation with continuous bathroom or central area **exhaust**.
- Manual balancing damper in outside air duct allows adjustment of the flow rate. Limit the outside air fraction to about 7% of total air handler flow (see **Figure 1** in Section 3.1.1).
- Periodic operation of the central air handler fan assures consistent ventilation air distribution and uniform air quality. It also reduces temperature variations between rooms.
- A motorized damper is required to close the opening to outside when the fan is off. With damper cycling control, outside air intake can be limited independently of how long the fan runs.
- Keeping all ducts inside insulated space provides the best performance. Sealed and well insulated ducts are next best.

7. Avoiding trouble by design

Experience is a great teacher, but much bad experience can be avoided through education. That is the goal of this section. Following some basic, uncomplicated design guidelines can go a long way to avoiding most trouble spots.

7.1 Ducts, fittings, grilles

A lot of time and effort can be spent following engineering procedures to design the ducts and fittings for a ventilation system. The best known of these procedures is the Air Conditioning Contractors of America (ACCA) Manual D. However, that level of detail is usually not needed for residential ventilation systems. Following are a few simple, common sense rules that will serve you well for most situations:

- Make changes in airflow direction as infrequently and as smoothly as possible.
- Pull the inner liner of flex duct to its full extent to avoid the spiral "accordion" effect which causes a lot of flow resistance, or use smooth metal duct.
- Start with the size of the fan outlet/inlet connection. Use that size if the duct length is fairly short (less than 10 ft.) Increase the duct size 1 inch if the duct run is not long (less than 25 ft.) and there are few fittings (less than 3). Go up 2 inches in duct size if the duct run is long or there are many fittings. Size the wall cap or roof jack to match the final duct size.
- Seal all joints with long-lasting material. Duct mastic is best. Some UL 181 listed tapes can also work well on clean surfaces, but do not use cloth-backed tape.

The outlet connection for standard bathroom exhaust fans is usually 3 inch diameter. The better fans are 4 inch. Remote fans usually have 4, 6, or 8 inch diameter inlet or outlet connections. Smooth increaser/reducer fittings to change duct sizes are commonly available and inexpensive. Don't ignore the odd sizes of 5 inch and 7 inch diameter duct. While they are less commonly stocked, they are available, and can make achieving the right air flow much easier.

If you are designing the duct system, when laying out duct runs and sizes, plan for air velocity of:

- 750 ft/min or less for exhaust ducts after the fan
- 350 ft/min or less for exhaust ducts before the fan (also called pickups)
- 500 ft/min or less for supply ducts

This will help keep static pressure and noise down, while keeping throw and efficiency up. Throw and efficiency have to do with how well the air is injected into the room so that it mixes well with room air.

General conventions for branching round ducts are:

- One 5" duct branches to two 4" ducts
- One 6" duct branches to two 5" ducts
- One 8" duct branches to two 6" ducts

Exhaust fans in the same dwelling can share a common discharge duct, but each fan must have a back-draft damper to prevent movement of air from one fan back through another.

Stamped metal grilles are the least expensive, but offer no adjustability and are the least efficient. Aluminum rather than steel should be used where moisture will be present. Grilles with adjustable curved blades to turn the flow to suit the situation are worth the investment. Grilles with a means to adjust the volume of air flow can help with balancing, but there are limitations. Inline balancing dampers do a better job of adjusting air flow.

Many good plastic grilles are available in sizes suitable for ventilation ducting. The best grilles have a means to close down the air flow and have curved diffuser-type surfaces to spread the air out evenly with little noise.

7.2 Condensation and Mold in or on Ducts

Condensation occurs when air is humid and surfaces are cold. To be more specific, the temperature of the surface has to be at or below the dew point temperature of the air for condensation to occur.

A soda can that you take out of the refrigerator is about 36 °F. In summertime in air conditioned buildings, the room air dew point temperature is usually 55 °F or higher. Those conditions will cause condensation to form on the soda can. In wintertime, if room temperature is 68 °F and the room relative humidity is less than 30%, then the soda can temperature is slightly higher than the room dew point and condensation will not occur.

Think of the inside surface of your central air conditioning system supply ducts at the end of a cooling cycle like the surface of the soda can. Then think of humid ventilation air flowing through those ducts from a single-point supply ventilation fan or an HRV. The supply duct surfaces are about 50 to 55 °F, and the summertime ventilation air dew point is usually above 65 °F in humid climates. Those are conditions for condensation inside the central supply ducts. If that occurs over a sufficient length of time, mold will grow.

Therefore, it is not recommended to inject ventilation air from HRV's or other ventilation supply fans into the central supply ducts during humid cooling conditions. Supplying ventilation air into the return side of the central system ducts can also be risky because air can flow through the central air handling unit while it is off. It is more feasible if a mixing ratio of at least 3 parts inside air is mixed with 1 part outside air and the air is injected upstream of the central system filter.

The following ventilation ducts need to be insulated to avoid condensation:

- Any exhaust ventilation duct going through wintertime cold spaces (avoids condensation on the inside of the duct)
- Any supply ventilation duct within the conditioned space (avoids condensation on the inside of the duct in summer and on the outside of the duct in winter).

7.3 Short circuiting

Short circuiting of ventilation air occurs when ventilation air enters and leaves a space or duct before it has a chance to mix well enough with room air to do the job it was intended to do—that is, to adequately dilute pollutants.

The most common occurrence of short circuiting is with HRV's and ERV's that are connected to the return side only, or connected to both the supply and return sides of a central air handling system. That configuration is common practice because it reduces the initial cost relative to an independent, fully ducted system. For that configuration to work, the central system fan must be operated whenever the HRV/ERV is on. Use of the thermostat fan-on selector must not be relied upon to solve this problem. If the central fan is not interlocked with the HRV/ERV, then the ventilation system will simply recirculate with the outdoors, having no positive effect on indoor air quality until there is a thermostat demand for central fan operation for heating or cooling. Constant operation of the central fan will result in high electric energy consumption and poor humidity control during the cooling season in humid climates. Energy consumption can be reduced through use of a timer to reduce the operational hours of the ventilation system and central fan, and through use of an ECM central fan.

Another example of short circuiting would be where an exhaust fan located in a utility room off the garage draws most of its air from the garage and exhausts it to outdoors. That will occur if the path of least resistance for air moved by the exhaust fan is through leakage in the enclosure separating the garage and the utility room. A similar short-circuit scenario could be true for any single-point exhaust fan located in a closed room.

Short circuiting of ventilation air can also occur for any system where air is supplied in close proximity to where air is returned. Where outlet or inlet placement does not allow for much separation, be sure to use a grille style that will throw the supply air away from the return inlet .

7.4 Lack of ventilation air distribution

Lack of ventilation air distribution always occurs when short circuiting occurs. However, lack of air distribution may also occur due to lack of ducting or whole-house mixing.

The best way to assure proper ventilation air distribution is to first avoid short circuiting, then duct ventilation air to each room or operate the central system fan periodically to achieve uniform mixing among all rooms.

7.5 Uncomfortable air complaints

It is impossible to provide ventilation air exchange without moving air. Some people are more sensitive to air flow from ducts than others. An effective ventilation system will not cause uncomfortable air complaints. If a ventilation system annoys the occupants they will not use it. When the ventilation system is not used, problems related to building durability, occupant comfort, and occupant health may arise.

Exhaust air does not cause uncomfortable air complaints because the air flow where people could feel it is slow and diffuse.

Supply air, whether from a ventilation fan or a central HVAC system, can feel uncomfortable because of its temperature, humidity, speed, direction, and volume flow rate. Temperature and humidity should be addressed by tempering or conditioning the air before delivery. Speed, direction, and volume flow rate should be addressed by proper duct design, grille placement, and grille characteristics. Too much air, moving too fast over people feels uncomfortable even if it is at room temperature. If the air is colder than room conditions, and if the people are sedentary (such as in bed) even more care must be taken to avoid direct contact. Uncomfortable direct contact is best avoided by:

- supplying smaller amounts of air to more locations;
- avoiding grille placement directly over locations where sedentary people will likely be; and
- using grilles that cause supply air to quickly mix with surrounding room air.

The following guidance is specific to ventilation systems that periodically operate the central air handler for ventilation air distribution and whole-house mixing:

- Educate the customer about the purpose of periodic air handler fan operation and its importance to whole-house ventilation air distribution, and its corollary benefits of more uniform comfort conditions.
- Do not feed a supply register with a duct larger than 8" diameter, except possibly for high ceilings in open areas.
- Do not feed a supply register in a bedroom with a duct larger than 6" and keep the duct air velocity below 500 ft/min, which will keep the volume flow below 100 cfm.

- Place registers so as to avoid blowing air directly on beds. Two-way and three-way registers will also help divert air evenly without concentrating the flow in one direction. This is especially important for master bedrooms. For example, use two supply registers in the master bedroom as opposed to one large register; always split an 8" duct into two 6" ducts for master bedrooms.

7.6 Where does the outside air come from?

Since a basic assumption of whole-house ventilation is that the more contaminated inside air is diluted with less contaminated outside air, it is important that the intake of ventilation air be from a known fresh air location. This is a disadvantage of exhaust ventilation and a benefit of supply and balanced ventilation.

With exhaust ventilation, the source of outside air cannot be known because it simply enters through the building enclosure based on the path of least resistance, which is constantly changing due to environmental forces of wind and stack effect. The ventilation air may come through a variety of building enclosure penetrations connected to the garage, crawlspace, attic, or foundation gaps, all of which would likely add to the indoor pollutant level. As air comes inward through the building enclosure, whatever it comes in contact with can have an effect on its contaminant level. The building materials may release VOC's and particulates, or have surface mold, and foreign pollutants may exist within the walls.

Supply or balanced ventilation on the other hand allows the point of outside air intake to be planned. The International Residential Code prohibits sources of outside air as follows:

Outside or return air for a forced-air heating or cooling system shall not be taken from the following locations:

1. ***Closer than 10 feet from an appliance vent outlet, a vent opening from a plumbing drainage system or the discharge outlet of an exhaust fan, unless the outlet is 3 feet above the outside air inlet.***
2. ***Where there is the presence of flammable vapors; or where located less than 10 feet above the surface of any abutting public way or driveway; or where located at grade level by a sidewalk, street, alley or driveway.***

That covers a lot of the potential for problems, but it is best to simply avoid locating an outside air intake on the roof for the following reasons:

- Air from exhausts, vents, and chimneys can too easily be re-entrained in outside air intakes due to the potential for mistakes in locating air intakes far enough away from outlets, as well as due to wind effects.
- Air intakes on shingle roofs can draw in asphalt fumes or odors.
- In summer, air taken from roofs is usually hotter than air taken from sidewalls.
- The additional roof penetration is another potential water leak.

Therefore, it is not recommended to take ventilation air from the roof. Taking ventilation air from a sidewall, gable end or soffit area is better. For 2-story homes, going out the band joist often works well. For 1-story homes that have no gable, a small 45 degree fur-down in a closet on an outside wall, or in the back corner of the garage works well to get to the sidewall below the cornice.

While a back porch ceiling/soffit can be a convenient place to locate an outside air intake, care should be taken so that outdoor cooking smoke is not likely to be drawn in.

Running ventilation ducts through the eave soffit is not recommended. If the outside air intake or exhaust outlet are run through the eave soffit instead of through the sidewall, specific care must be taken to:

1) not crush the ventilation ducts between the wall top plate and the roof sheathing;

2) leave the full depth of ceiling insulation in place; and

3) fully terminate the ducts through the finished soffit to outdoors so that air flow is not restricted and air leakage to/from the attic will not occur.

7.7 Fan flow ratings

Manufacturers usually give a rated air flow based on duct system static pressure. The higher the static pressure, the lower the air flow. The two rating points are 0.1 inch water column (w.c.) and 0.25 inch w.c. In Pascals (Pa), those points are 25 Pa and 62.5 Pa.

Properly designed and installed systems should be able to keep the static pressure below 0.25 inch w.c. However, it is not uncommon to find systems in the field at 0.4 to 0.6 inch w.c. (100 to 150 Pa). Depending on the specific fan and its ability to handle high static pressure, such systems may have insufficient air flow.

7.8 Fan flow verification

After installation, actual air flow should be verified by testing and compared to the design air flow. The most common test instrument for measuring volumetric flow (cfm) is a flow hood. The flow hood is placed over the supply or return grille and a calibrated velocity grid inside the hood measures the air flow.

A calibrated fan that is commonly used to test ducts for air leakage can be used to provide the most accurate ventilation air flow measurements. A box connected to the calibrated fan is placed over the supply or return grille and the pressure differential between the box and the room is taken to zero by increasing the fan speed. When the pressure difference is zero, the measured air flow through the calibrated fan is representative of the air flow due to the ventilation fan.

Flow grids, also called flow stations, can be used inline with ducts. By measuring the pressure differential across the flow station, the air flow can be calculated using a chart or equation given by the manufacturer. These devices usually require some straight duct distance upstream and downstream to give reliable results.

Other handheld devices can measure air velocity in a duct or at the face of a grille, but obtaining the average velocity over the entire cross section is difficult and more uncertain.

7.9 Noise

Sound levels for ventilation fans are measured in sones. HVI literature states that the sound level of 1 sone is about the same as a quiet refrigerator running in a quiet room. Fans used for ventilation should be 1 sone or less, unless they are remotely mounted, or part of the central air distribution system. Fans that are too noisy cause complaints or get shut off.

Even a low-sone fan will become noisy if ducted improperly. Improper ducting results in excessive air turbulence and/or high static pressure. Several simple rules for reducing fan and air flow noise are:

- Keep the duct system static pressure below 0.3 inch w.c.
- Don't change the air flow direction on the discharge side of the fan for at least 18 inches after the fan.

- Where ducting for remote fans is short (less than 8 ft), use flex duct.
- Hang remote fans with metal straps or clips where possible. Otherwise, use sound/vibration dampening material underneath mounting points if fastened to or resting on framing.

7.10 Maintenance

Maintenance must be easy or it won't get done. Clogged air filters are probably the most common maintenance failure. Air filters must be easy to access. They should be either washable or readily available to purchase, preferably at the common home center stores. Outside air intakes that go through the first floor band joist are items that require annual cleaning. Avoid placing outside air intakes less than 12 inches or so off the ground. Parts that are expected to wear out and need replacement, like drive belts or moisture transfer cores, often don't get noticed when broken, or replaced when needed. Homeowners are usually less aware of maintenance needs for ventilation systems that are not part of the central space conditioning system. If the central space conditioning system fan stops, the system will surely receive the needed attention. Education through verbal and written information is important to make sure that ventilation system maintenance is not ignored.

Appendix

Ventilation System Decision Flow Chart

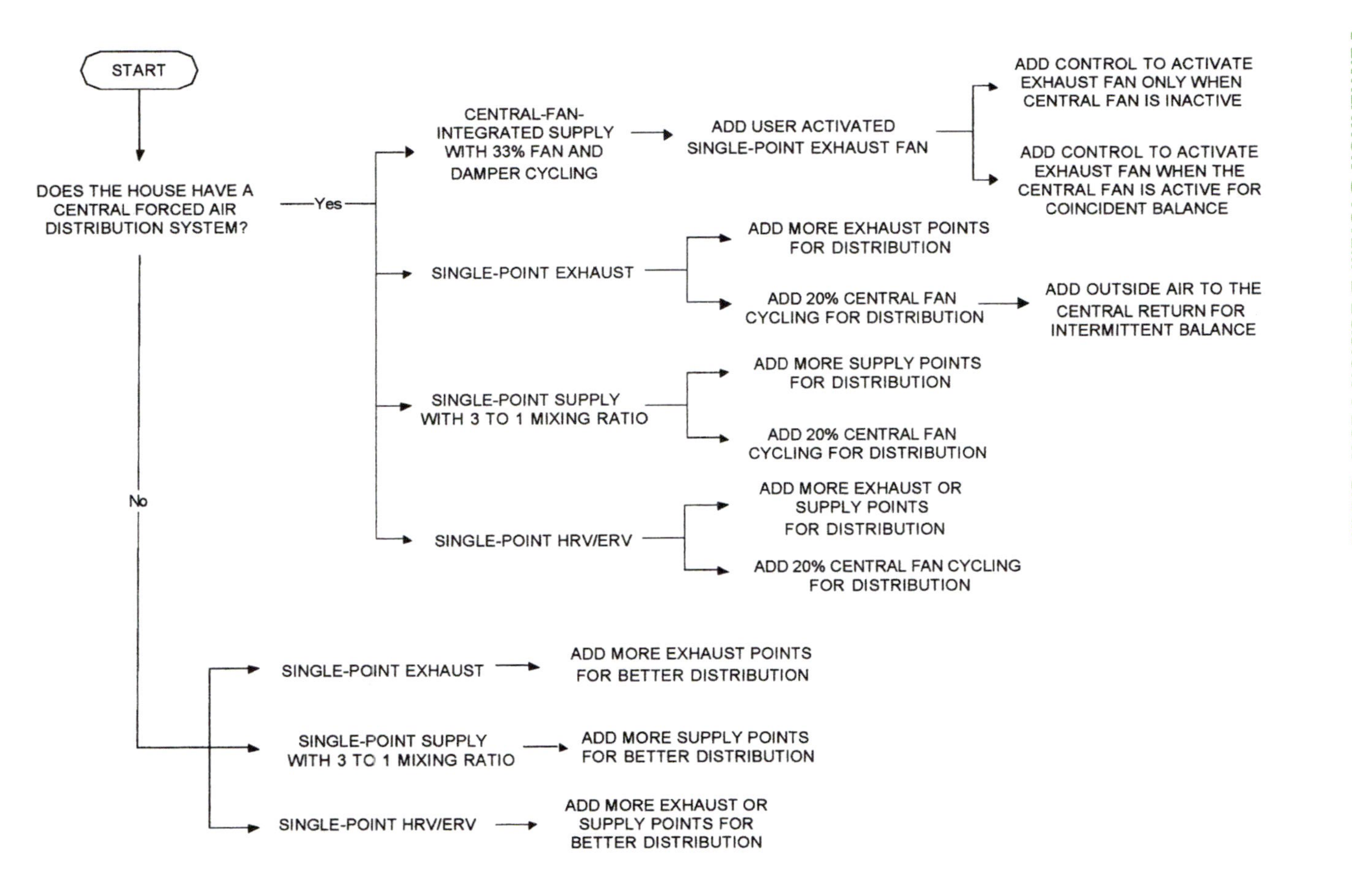

Notes